Wireless Positioning Technologies and Applications

For a listing of recent titles in the *Artech House GNSS Technology and Applications Series,* turn to the back of this book.

Wireless Positioning Technologies and Applications

Alan Bensky

ARTECH
HOUSE

BOSTON | LONDON
artechhouse.com

Library of Congress Cataloging-in-Publication Data
A catalog record for this book is available from the U.S. Library of Congress.

British Library Cataloguing in Publication Data
A catalogue record for this book is available from the British Library.

ISBN-13: 978-1-59693-130-5

Cover design by Igor Valdman

© 2008 ARTECH HOUSE, INC.
685 Canton Street
Norwood, MA 02062

10 9 8 7 6 5 4 3 2 1

To my wife Nurit and our lovely daughters, Chani, Racheli, and Ortal

Contents

Preface   *xiii*

CHAPTER 1
Introduction   1
1.1   Fundamentals and Terms   1
    1.1.1   Basic Measurements   2
    1.1.2   Terms   3
1.2   Applications   4
    1.2.1   Cellular Networks   4
    1.2.2   Person and Asset Tracking   5
    1.2.3   Wireless Network Security   6
    1.2.4   Location-Based Advertising   6
    1.2.5   Location Services for Vehicles and Traffic   6
1.3   Overview of Distance Measurement and Location Methods   6
1.4   Organization of the Book   10
    References   11

CHAPTER 2
Basic Principles and Applications   13
2.1   Signal Parameters   14
    2.1.1   Time Resolution   14
    2.1.2   Pulse Width and Duty Cycle   15
    2.1.3   Bandwidth   15
    2.1.4   Noise   17
    2.1.5   Pulse Compression   20
2.2   Basics of Location   27
    2.2.1   Rho-Theta   27
    2.2.2   Theta-Theta or AOA   28
    2.2.3   Rho-Rho or TOA   29
    2.2.4   TDOA and Hyperbolic Curves   30
2.3   Navigation Systems   32
    2.3.1   DME   33
    2.3.2   VOR   37
    2.3.3   Loran-C   38
    2.3.4   GPS   44
2.4   Conclusions   51
    References   52

CHAPTER 3

Spread Spectrum   53

3.1   Principles of Direct Sequence Spread Spectrum   53
      3.1.1   Transmitter and Receiver Configurations for DSSS   54
      3.1.2   DSSS Waveforms   55
      3.1.3   Despreading and Correlation   57
      3.1.4   Code Sequence Generation   59
      3.1.5   Synchronization   60
      3.1.6   Velocity Estimation   62
3.2   Acquisition   63
      3.2.1   Code Acquisition   64
      3.2.2   Carrier Acquisition   69
      3.2.3   Code Rate Matching   70
      3.2.4   Effect of Data Modulation on Acquisition   71
      3.2.5   Radiometric Detection   71
3.3   Tracking   72
      3.3.1   Carrier Tracking   72
      3.3.2   Code Tracking   73
3.4   Measurement of Elapsed Time   77
      3.4.1   One-Way Systems   78
      3.4.2   Two-Way Systems   78
      3.4.3   The Time Measurement Process   79
      3.4.4   High-Resolution Elapsed Time-Measuring Receiver   82
      3.4.5   Duplex and Half Duplex Two-Way Ranging Examples   83
      3.4.6   Sequence Length and Chip Period   87
3.5   Propagation Time Resolution   88
      3.5.1   Tracking Accuracy and Noise   88
      3.5.2   Multipath   90
      3.5.3   Increased Range Resolution Using Carrier Phase   92
3.6   Conclusions   93
      References   94

CHAPTER 4

Time Transfer   95

4.1   Time Transfer Basics   95
4.2   Calibration Constants   97
4.3   Range Uncertainty   98
      4.3.1   Clock Drift and Measurement Time   99
      4.3.2   Noise   101
      4.3.3   Multipath   102
      4.3.4   Relative Motion   102
4.4   Ranging Procedure in Wireless Network   103
4.5   Conclusions   105
      References   105

CHAPTER 5

Multicarrier Phase Measurement   107

5.1 Principle of Multicarrier Phase Measurement   107
5.2 Phase Slope Method   108
5.3 Phase Error Versus Signal-to-Noise Ratio   111
5.4 Estimation of Distance Variance Versus SNR   115
5.5 Multipath   118
5.6 System Implementation   123
     5.6.1 Phase Difference Measurements and Analogy to TDOA   125
5.7 OFDM   126
     5.7.1 The Basics of OFDM   126
     5.7.2 OFDM Distance Measurement   130
     5.7.3 Location Based on OFDM Distance Measurement   134
     5.7.4 Resolution of OFDM Distance Measurement   136
5.8 Conclusions   137
     References   138

CHAPTER 6

Received Signal Strength   139

6.1 Advantages and Problems in RSS Location   139
6.2 Propagation Laws   140
     6.2.1 Free Space   140
     6.2.2 Free-Space dB   140
     6.2.3 Open Field   141
     6.2.4 Logarithmic Approximation   142
     6.2.5 Randomizing Term X   143
     6.2.6 Outdoor Area Networks   144
     6.2.7 Path Loss and Received Signal Strength   146
6.3 RSS Location Methods   146
     6.3.1 RSS Location from Range Estimations   146
     6.3.2 RSS Location Based on Database Comparison   147
6.4 Conclusions   158
     References   158

CHAPTER 7

Time of Arrival and Time Difference of Arrival   161

7.1 TOA Location Method   162
     7.1.1 Overdetermined TOA Equation Solution   163
     7.1.2 TOA Method in GPS Positioning   166
7.2 TDOA   170
     7.2.1 TDOA Measurement Techniques   171
     7.2.2 Multilateral and Unilateral Topologies for TDOA   173
     7.2.3 TDOA Geometric Model   175
     7.2.4 TDOA Example   177

7.3   Performance Impairment   181
      7.3.1   Uncertainties in Data Measurement   181
      7.3.2   Random Noise   182
      7.3.3   Dilution of Precision (DOP)   182
      7.3.4   Multipath   184
      7.3.5   Cochannel Interference   186
7.4   Conclusions   186
      References   187

CHAPTER 8
Angle of Arrival   189

8.1   Triangulation   189
8.2   Antenna Performance Terms and Definitions   191
8.3   Finding Direction from Antenna Patterns   194
8.4   Direction-Finding Methods   198
      8.4.1   Amplitude Comparison   198
      8.4.2   Phase Interferometer   200
8.5   Electronically Steerable Beam Antennas   207
8.6   ESPAR Antenna Array   214
8.7   Conclusions   220
      References   221

CHAPTER 9
Cellular Networks   223

9.1   Cellular Location-Based Services   223
9.2   Cellular Network Fundamentals   224
      9.2.1   GSM Transmissions   226
      9.2.2   CDMA   227
      9.2.3   UMTS   228
9.3   Categories of Location Systems   229
9.4   GPS Solution   230
9.5   Cell-ID   231
9.6   Location Technologies Using TDOA   232
      9.6.1   Enhanced Observed Time Differences   234
      9.6.2   Observed Time Difference of Arrival   235
      9.6.3   Uplink Time Difference of Arrival   236
9.7   Angle of Arrival   236
9.8   Received Signal Strength and Pattern Recognition   236
9.9   Problems and Solutions in Cellular Network Positioning   237
      9.9.1   Narrowband Networks   237
      9.9.2   CDMA   237
      9.9.3   GSM   238
9.10 Handset-Based Versus Network-Based Systems   238
9.11 Accuracy Factors   239
9.12 Conclusions   239
      References   240

CHAPTER 10

Short-Range Wireless Networks and RFID 241

10.1 WLAN/WiFi 242
    10.1.1 TOA 242
    10.1.2 TDOA Methods for WLAN Location 248
    10.1.3 Fingerprinting 249
10.2 WPAN 251
    10.2.1 Bluetooth 251
    10.2.2 ZigBee 255
    10.2.3 Alternate Low Rate WPAN Physical Layer IEEE 802.15.4a 257
    10.2.4 ECMA-368 Standard 258
10.3 RFID 259
    10.3.1 Proximity Location 260
    10.3.2 Distance Bounding for Security 260
    10.3.3 Accurate RFID Location 262
10.4 Conclusions 262
    References 263

CHAPTER 11

Ultrawideband (UWB) 265

11.1 Telecommunication Authority Regulations 265
    11.1.1 FCC Regulations 265
    11.1.2 UWB in the European Community 267
11.2 UWB Implementation 268
    11.2.1 Impulse Radio UWB 268
    11.2.2 OFDM 272
11.3 IEEE 802.15.4a 274
    11.3.1 Physical Layer Characteristics and Synchronization 274
    11.3.2 Ranging Protocol 280
11.4 Dealing with Multipath and Nonline of Sight 283
    11.4.1 Multipath 283
    11.4.2 Nonline of Sight 284
11.5 Conclusions 286
    References 286

Bibliography 289

List of Acronyms and Abbreviations 293

About the Author 297

Index 299

Preface

This book is about wireless position estimation—how it works and what it's used for. Range and location are two aspects of position that are related geometrically. Most applications of distance measurement are aimed at determining a radio terminal's location. The book explains the relationships between the two concepts.

In some wireless systems, distance measurement and position location capabilities have become a necessary adjunct to communication. A big thrust in this direction occurred in 1995 when the FCC issued its directive for expanding 911 caller location services to cellular telephony. Other applications where position location is gaining importance are RFID, WLAN, and WPAN. Probably the best known and most widespread example of wireless location is the global positioning system (GPS). While several excellent sources describe GPS in great detail, this book explains the basics of the system and uses it as an example to demonstrate the application of fundamental distance measuring principles.

The first two chapters of the book describe typical applications and give a basic description of positioning methods, as well as definitions of important parameters and physical limitations of time measurements. The starting point for describing wireless distance measurement is radar. Its underlying principal is the determination of the propagation time and direction of a radio wave bounced back from a distant passive target. In contrast to the radar concept, this book deals with distance measuring between two or more active wireless terminals. An early example is the aeronautic DME, which originated in World War II and is still in use today. A related instrument is VOR—a wireless technology for obtaining bearings. Using DME and VOR together, a pilot can determine his location. A third system, Loran-C provides ships and planes with their geographic coordinates over large surfaces of the Earth. While these mature navigation aids are gradually being replaced by GPS, they serve as concrete examples of implementation of the basic positioning methods described in the book.

Subsequent chapters are organized in three broad classifications: technology, methods, and application. *Technologies* are the underlying communication systems and measuring techniques, including for example spread spectrum and ultrawideband, whereas *methods* relate to the physical and geometric principles, such as time of flight of the radio signal, received signal strength, and angle of arrival. Important and widespread applications are cellular handset location, short-range wireless networks, and radio frequency identification.

Along with the theoretic basis of each technology, the book has practical information on implementation. It discusses the impairments to achieving theoretical accuracy due to noise, multipath, and fading, and practical limitations of antenna directivity and time measurement precision. In addition to thorough coverage of the prominent technologies in use today, the book concludes with a description of ultrawideband, one of the most important directions of development in the near future.

This book is primarily aimed at working engineers who are assigned to projects involving wireless distance measurement and location or who want to expand their knowledge of wireless services. An understanding of basic engineering mathematics, including familiarity with Fourier analysis, matrix manipulation, and introductory probability, will be helpful for understanding some of the equations and examples.

Underlining the increasing interest in combining location awareness with communications is the fact that new industry standards are including specifications for ranging capabilities. Two examples are the ECMA-368 specification for high rate personal area networks and IEEE 802.15.4a, which extends the capabilities of the physical layer of the protocol known as ZigBee for low-cost, low-complexity sensor networks. These are described in Chapters 10 and 11, respectively. The inclusion in one volume of the legacy navigation systems and the ranging features in the newest specifications demonstrates the author's objective of providing a comprehensive in-depth review of distance measurement and location technologies.

Considering the interest in ranging and positioning for a variety of wireless communication applications, there are surprisingly few books that cover the wide range of technical approaches to the subject. Generally, these books deal separately with specialized applications or technical aspects of distance measurement and location. Much information is available in technical articles published in professional journals. These articles often describe academic investigations at a level that is not accessible to many technical people involved in product development. Sources of advanced studies are cited in the text and in the bibliography of this book, so that anyone who wants to delve deeper into a particular topic will know where to obtain additional material. However, a person working on a project involving distance measurement or location needs a source from which he or she can orient themselves on the various technologies—the theory of operation, examples of implementation, advantages and disadvantages—and to help them assess the applicability to what they aim to accomplish. The fulfillment of that need, then, is the purpose of this book.

CHAPTER 1

Introduction

The basic task of a wireless communication system is to transfer information originating at one terminal to one or more other terminals. However, by using characteristics of the transmitted signal itself, another use has been added for wireless systems. It involves estimating how far one terminal is from another, or where that terminal is located. The uses for wireless distance measurement and location is varied and their numbers are constantly growing. They are included in areas of personal safety, industrial monitoring and control, and a myriad of commercial applications. The methods used for getting location information from a wireless link are also varied. Complexity, accuracy, and environment, are among the factors that play a role in determining the type of distance measuring system to apply for a particular use, although there may be several competing methods employed for the same type of application.

Perhaps the epitome of wireless location is the global positioning system (GPS). Although this book covers briefly the workings of that system, most of it is concerned with wholly on Earth technologies. GPS remains a prime example of the relatively high positioning accuracy obtainable through very long distance wireless links. However, it is not a solution for everyone. GPS is well-priced for some applications, but overly expensive and complex for others. Also, GPS performance deteriorates in indoor use and in urban environments.

1.1 Fundamentals and Terms

Distance measuring and location are two closely related concepts. Distance measuring can be considered as determining the radius of a circle or of a sphere. Location is a point in space which is described symbolically or as a set of coordinates defined as distances or angles in relation to another point either locally or globally, say, in terrestrial three-dimensional coordinates. Distances may be used to compute location, or knowing location distance can be found. Most applications want to pinpoint location but there are some for which distance itself is adequate. For example, knowing the distance of a station from a wireless LAN access point may be enough to determine if it is a legitimate member of a network and can be allowed to log in. As another example, the closeness of a purchaser to a checkout counter in a supermarket or to a gas pump may be suitable for deciding whose cell phone to bill.

1.1.1 Basic Measurements

There are three basic properties that enable distance measurement and location from analysis of specific physical characteristics of radio signals:

1. *Received signal strength (RSS)*. The power density of an electromagnetic wave is proportional to transmitted power and inversely proportional to the square of the distance to the source. This physical law as well as the vectorial combination of waves that reach a receiver over different paths are the basis for estimating distance and location from signal strength measurements.

2. *Time of flight (TOF)*. The distance between a transmitter and receiver equals the time of flight, or electromagnetic propagation time, of the transmitted signal times the speed of propagation, which is the speed of light. Distance can be determined from measurement of time of arrival (TOA) of a signal at a receiver when transmission time is known, or from differences of reception time at different locations (time difference of arrival—TDOA). Another expression of time of flight is the phase of the received signal, which may be referred to as phase of arrival (POA), since phase may be related to time and distance through the signal wavelength and speed of light [1].

3. *Angle of arrival (AOA) or direction of arrival (DOA)*. The wavefront of a transmitted signal is perpendicular to the direction of propagation of the wave. The direction of a radio wave can be estimated by varying the known spatial radiation pattern of the transmitting or receiving antenna while noting the change in received signal strength. The angle of arrival may be determined to be the point in the pattern rotation where the signal strength is maximum, or where a null in signal strength occurs, depending on the design point of reference. In contrast to the RSS method, knowledge of transmitted power is not required. Distance cannot be found directly using an AOA measurement. At least two AOA measurements or an AOA measurement and TOF or RSS measurements are required to determine the position of a wireless terminal.

All methods of distance measurement and location are derived from the measurements described above, alone or in combinations. There are two variants of these methods that differ enough from the normal measurement case so that they may placed in separate classes:

1. *Proximity* refers to detection of a mobile terminal as being within radio range of a fixed location so that the mobile is known to be within an area around that location.

2. *Fingerprinting* locates a terminal by comparing various characteristics of a signal or signals received at or from that terminal with a database of the same type of characteristics that has been compiled in advance over a given area or volume.

Velocity is often estimated as an adjunct to distance or location. Relative speed between two terminals is calculated as the change in distance divided by the time difference between two distance measurements. Similarly, velocity of a terminal is found from the vectorial difference of two sets of coordinates of the moving terminal divided by the difference in measurement times of these sets of coordinates. When a terminal is moving, or a velocity measurement is required, the time required to make a measurement is significant. Relative velocity can also be determined by measuring the Doppler shift of a carrier frequency.

1.1.2 Terms

Many of the terms that are used throughout this book are defined below. Some have similar meanings, and although there may be small distinctions between them, they are considered synonyms when using them interchangeably will not cause a misunderstanding. For example, location and position are often considered to mean the same thing.

Location answers the question "Where is it?" It may be a symbolic place, like a room or street, or defined by coordinates in two or three dimensions. *Position*, as used in this book, is generally synonymous with location but may also refer to location and distance from a terminal collectively. *Positioning* usually refers to the process of finding the two- or three-dimensional coordinates of a terminal but can mean the determination of distance or range. A position may be relative or absolute. A relative position is described by a distance or bearing in relation to a particular object. An absolute position has two- or three-dimension coordinates that are common to a large defined region (e.g., the globe).

Range has two meanings. Most correctly, it is the greatest distance between two terminals over which communication is supported. It also may be interpreted as any distance between terminals that has been calculated by distance measurement or location techniques.

Three common terms classify the geometric procedure for finding location from distance or angle of arrival measurements. We will refer in general to the calculation of a two- or three-dimensional position from unilateral or multilateral measurements (defined below) as *triangulation*, since the minimal form of base station/ mobile station layout is a triangle. To be more specific, location determination from multiple distance measurements is called *lateration*, while *angulation* refers to the use of angle or bearing data relative to points of known position to find a target's location [2]. *Trilateration* is the word commonly used when position is derived from the measured or given lengths of the three sides of the triangle.

Terminal or *station* refers to either side of a communication link. It may contain a transmitter, receiver, or transceiver. A fixed terminal is a terminal whose position or location is known. A base station is a fixed terminal. The coordinates of a mobile terminal or mobile station (MS) are generally, but not always, unknown.

Target refers to a terminal whose location is to be determined. It is generally mobile.

A *beacon* is a continuous or periodic transmission that facilitates timing synchronization or position measurements between terminals.

Location systems may be classed as multilateral or unilateral [3]. In a *multilateral* system the target is a transmitter whose location is calculated from measurements taken by multiple fixed terminal receivers whose positions are known. An example is cellular positioning where several base stations time the reception of a handset's transmission and the network performs the position estimate. By contrast, the target in a *unilateral* system receives transmissions from multiple terminals, whose positions are known, and calculates its own position. GPS is a unilateral system. The multiple transmitting satellites are in motion but their exact positions at the time of transmission can be calculated by the target receiver, which determines its own location. Loran-C is another example of a unilateral system.

Epoch is a particular instant on the baseband waveform, for instance the start of a particular frame or the first bit after a synchronizing preamble. It also refers to an interval, identified by chip number, relative to the beginning of a pseudorandom noise sequence.

Time of flight (TOF) is the time interval between transmission time of an epoch to its reception at a distant receiver. The term is used in this book to designate the distance measuring method that is based on the propagation time of an electromagnetic wave.

Navigation is the determination of the position and velocity of a moving vehicle [4].

Accuracy refers to how closely a measured distance agrees with the actual distance.

Precision is an indication of the repeatability of a measurement [2]. *Resolution* corresponds to the markings on a ruler—graduations could be in millimeters or tenths of a millimeter, for example. As another example, a digital clock readout may indicate time intervals with resolution as good as a hundredth of a second, but it would have poor accuracy if the time indicated differs from the actual time by several seconds. Often precision and resolution are used as synonyms.

1.2 Applications

A few of the myriad applications of distance measurement and location are described below. They illustrate the various technologies and somewhat the degree to which this ancillary communication service has penetrated divergent activities. Wireless location applications are often referred to as location-based services (LBS), particularly as provided by cellular networks for adding value to their basic service of mobile telephony [5].

1.2.1 Cellular Networks

The impetus for providing location ability on cellular networks was triggered by the requirement of the Federal Communications Commission (FCC) for position information on E-911 emergency cell phone calls. The European equivalent call number is E-112. There are basically two methods of operation of cellular location. In one, there are no changes to the handset. Two or more base stations note time of arrival of signals from the handset, and by way of triangulation determine its location. The other method involves a location determination facility in the handset. The location method used may be based on the network itself or on an independent system. For example, a GPS unit in the handset sends its position over the cellular network. Various commercial location-based services utilize the position-finding feature of cellular networks [5].

Location awareness also can provide information to augment network performance and efficiency. Independent location determination in the handset can be used to help the network with handoff from one base station to another. If the network knows where the handset is located at all times, it can decide when to handoff to a different base station. Obtaining information on the location of subscribers relative to the various cells can help to plan system loading and channel allocation as well as the deployment of additional cells [6].

1.2.2   Person and Asset Tracking

Many wireless location systems have been developed for varied use in hospitals. Some uses involve finding the whereabouts of hospital equipment, which have radio frequency tags attached to them. Similarly, staff members can be located at any time without their intervention. Tags are also attached to newborn babies to prevent them from being abducted.

There are several systems operating that are intended to prevent vehicle theft and to locate the vehicle after it has been stolen. When triggered, a transceiver located in the vehicle sends periodic beacons to fixed units that are part of a regionwide network. A control hub tracks the vehicle and notifies authorities where it is located.

Tagging pieces in inventory is a common use of location systems. RFID tags, normally active, attached to items in a warehouse can be located through any one of the methods used for wireless location.

The problem of a child going astray in a large amusement park, for example, can be solved by giving the child a transceiver mounted on his or her hand like a watch. The child's movements anywhere in the park can be monitored by a central station that is accessible by the parents.

Wireless location systems are used extensively for tracking wildlife. Animals are captured and after battery-operated transmitters are attached to them, they are released to their normal habitat. The approximate location of the animals can be tracked by portable receivers and directional antennas. Special tracking satellites are sometimes used. GPS receivers that are included in the transmitter can detect the animal's position and send coordinate data to a satellite or a receiving terminal in the vicinity [7].

Wireless handcuffs are used to confine criminals to their residence instead of taking up space in a penal institution. A monitoring device triggers an alarm if the criminal leaves the area where he or she is allowed to be. Operating on a similar principle, body-mounted transmitters on Alzheimer patients detect when the person goes astray and summon help [6].

1.2.3 Wireless Network Security

Wireless networks use location methods to enhance security. A rogue terminal can be discovered to be outside the premises of an office, for example, and access to the network can then be prevented, or the terminal can be located.

1.2.4 Location-Based Advertising

Location methods can be used for selective advertising. The cell phone of a visitor to a mall can be located and can be made to display an advertisement of a shop very near to where its holder is located [5].

1.2.5 Location Services for Vehicles and Traffic

Dedicated short-range communication (DSRC) refers to wireless protocols specifically designed for automotive vehicles. Operating in frequency bands between 5 and 6 GHz in several regions, including the United States, Europe, and Japan, the technology aims to enhance the safety and the productivity of the transportation system. Among its applications, pertaining to wireless location, are intersection collision avoidance and electronic parking payments.

A term for describing traffic-related services is traffic telematics [5]. Several of the services are related to wireless location. A transponder mounted in a vehicle is signaled when the vehicle travels on a toll road and the owner can be automatically billed without the necessity of a toll station. Another use is monitoring the kilometrage of vehicles on highways for the purpose of determining road usage fees. Such devices are principally based on proximity and do not directly measure distance or position. Other services are fleet management and mobile marketing, based on location relative to, for example, banks, restaurants, and tourist attractions that use the technology to selectively advertise their wares.

1.3 Overview of Distance Measurement and Location Methods

This section contains an overview of what distance measurement is and its relationship to location.

The classic and best-known wireless distance measuring technology is radar, originally an acronym for radio detection and ranging. In its simplest form, it works by transmitting a radio frequency pulse from a directional antenna and measuring the elapsed time until the pulse, reflected from a target object, is detected by the radar receiver. Transmitted and reflected pulses are shown in Figure 1.1. The distance to the target, or range, is calculated by multiplying the elapsed time, τ, by the speed of radio wave propagation, which is the speed of light, and dividing the result by two. The target position is found from noting the direction the radar antenna is pointing at the time the reflected pulse is received, which can be expressed in three dimensions by the azimuth and elevation angles of the antenna relative to a given coordinate system and the range. Rectangular, cylindrical, and spherical coordinate systems are illustrated in Figure 1.2.

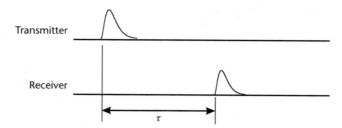

Figure 1.1   Transmitted and received radar pulse.

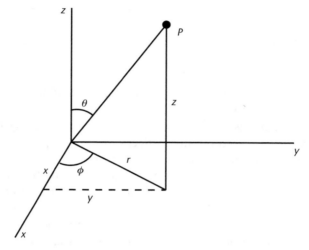

Figure 1.2   Coordinate systems: Rectangular coordinates (x, y, z), cylindrical coordinates (r, θ, z), and spherical coordinates (r, θ, φ).

In this book, the systems used for distance measurement and location involve at least two active terminals. We can extend the radar example to include two active terminals by using what is called a transponder. Instead of being passively reflected back, the radar pulse that is received at the target is retransmitted by the transponder. The radar receiver determines distance to the target by measuring the time delay between transmission and reception but in this case the signal from the transponder is much stronger than that received after passive reflection from a target. This time delay is referred to as time of flight. In order for that time delay to be equal to the two-way signal travel time, the transponder would have to resend the signal immediately upon its reception. This is possible only if the transmissions in each direction are on frequency channels that are separated enough so that simultaneous transmission and reception are possible without self-interference. If only one channel is used for communication in both directions, then retransmission at the transponder can begin only after the initiating terminal has stopped sending and the transponder has time to change its mode from receiver to transmitter. This is shown in Figure 1.3. If the waiting time *T1* in Figure 1.3 is constant and known to the initiator, it can be subtracted out and the distance between initiator and target can be calculated in the same manner as for passive reflection from a target in a radar system.

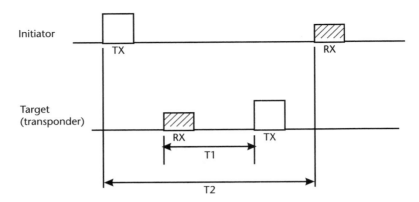

Figure 1.3 Initiator-transponder pulse timing.

Radar determines location using only two terminals by a combination of TOA and AOA. When only a distance measurement between two terminals is required, a directional antenna is not needed. In the case of the radar system, an omnidirectional antenna cannot be used because reflections from objects in all directions would be received and the target could not be identified. Also, the directional radar antenna has high gain, which increases the signal-to-noise ratio of the received echo. However, using an active target only its known signal would be considered by the initiator's receiver to make the distance calculation. In most of the applications we have looked at previously, position, or location, is needed, but it is not practical to use mechanically or electronically steerable direction antennas due to size, cost, or accuracy considerations. If nondirectional antennas are used, accurate target location is possible through the use of multiple, spaced terminals. An example is shown in Figure 1.4. A, B, and D are fixed terminals whose coordinates in a given reference system are known. C is a mobile terminal whose location is to be determined. In this example all four terminals are in the same plane. Distances AC and BC can be found by measuring the time of flight as described previously.

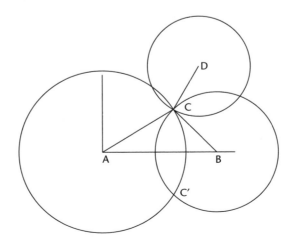

Figure 1.4 Positioning using intersecting circles.

Considering each of the terminal pairs A-C and B-C, C lies on the perimeter of a circle whose center is the fixed terminal and radius is the measured distance. The intersection of the circles centered at A and B is the location of C. Figure 1.4 shows there is an ambiguity in the determination of C's position when calculated using only the two fixed stations A and B. A false position, at C′, is also an intersection of the two circles. The true position of the mobile terminal can be known only from additional information about its whereabouts, or by measuring distance to an additional terminal, D. In Figure 1.4, C is at the point of intersection of the circles whose radii are three distance measurements.

With the AOA method, target location is the point of intersection of bearing lines between fixed stations and the target [8]. Steerable unidirectional antennas are situated at two fixed terminals whose coordinates are known, as illustrated in Figure 1.5. The angle of arrival of signals transmitted from the mobile target terminal C at each of the fixed terminals A and B, and the distance AB found from the known coordinates of A and B, are used to calculate the coordinates of T.

There is another way to estimate distance between two radio terminals without using the methods of finding time of flight or angle of signal arrival as described above. It involves determining range from the strength of a received signal. Signal strength generally decreases as distance between terminals increases according to a law whose details depend on signal path propagation in the physical environment where the terminals are located. In free space the propagation is such that the received signal voltage is inversely proportional to the distance between the transmitter and the receiver. In other than free space this law is modified by reflections from nearby objects and obstructions in the signal path. Signal strength can be used as a basis for estimating distance if the propagation behavior is known and, additionally, if the transmitter power and actual transmitter and receiver antenna gains are known, as they are oriented during the measurement. In many circumstances the propagation law is not constant in the region of interest. Some location systems use a database of path loss throughout the relevant region in order to make distance approximations between terminals. Each fixed station could have its own database containing distance contours. Location of a mobile station can then be estimated from the intersection of contours determined at two or more fixed stations, similarly to the intersection of circles shown in Figure 1.4.

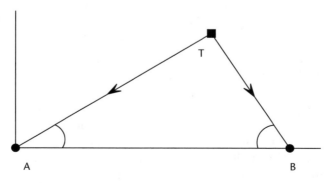

Figure 1.5   Angle of arrival positioning.

Related to the concept of signal strength distance measurement is the use of proximity to find location. A mobile terminal can get an approximation of its position within a network of fixed stations by determining through relative signal strength the fixed terminal to which it is closest. It then can consider its location as that of the nearest fixed station.

Probably the most influential wireless location method is GPS. It is based on measuring with great accuracy the distance between a mobile terminal to each of several members of a constellation of satellites and calculating from geometric relations the mobile's location relative to a universal coordinate system. Although the technology fits in the category of time of flight, GPS deserves a class for itself since the details of its operation are firmly fixed and it is available in complete modules that are integrated into products serving varied distance measurement and location applications. The location results from a GPS measurement can be presented in many forms to suit those applications, and auxiliary technologies allow enhancing accuracy and speed of reporting.

To sum up this section, we classify distance measurement and location technologies in the following categories:

- Time of flight;
- Angle of arrival;
- Signal strength;
- Proximity;
- Fingerprinting.

As mentioned, proximity and fingerprinting are really subcategories of signal strength and sometimes time of flight, but it is convenient in discussing and comparing application methods to refer to them separately.

1.4 Organization of the Book

In this chapter, some of the more common applications for wireless distance measurement and location were presented. The basic technologies were then described briefly as a preliminary introduction to the details of operation. Chapter 2 gives details of parameters involved and methods of implementation. It also gives examples of the use of the technologies in legacy navigation applications and GPS.

Chapters 3, 4, and 5 describe ways of finding the distance between two terminals by methods based on radio signal time of flight. Chapter 6 discusses the use of the signal strength method, which is generally much easier to implement than that of time of flight, but is in most cases not capable of similar accuracy. The geometric use of distance measurement to determine location is described in detail in Chapter 7. This chapter explains performance impairments and ways to overcome them.

Chapter 8 is about the angle of arrival method of wireless location. While generally of lesser use in short-range wireless location systems than the methods of time of flight, it does have particular potential advantages that may make it more important as electronically steered "smart" antenna arrays become more sophisticated and prevalent.

Chapter 9 discusses one of the most important areas for location awareness—cellular networks. While applications are fairly uniform, the chapter shows that there are varied technologies for implementing them.

Perhaps the fastest growing area of location awareness applications is in local and personal area networks. They are the subjects of Chapter 10. This chapter also includes RF identification (RFID), where the use of location applications is also rapidly expanding.

The book is completed in Chapter 11 with a description of principles and implementation of ultrawideband (UWB), which appears to be the most important communications technology now being adopted for high-accuracy short-range distance measuring and location applications.

References

[1]   Pahlavan, K., and X. Li, "Indoor Geolocation Science and Technology," *IEEE Communications Magazine*, February 2002.

[2]   Hightower, J., and B. Borriello, "Location Systems for Ubiquitous Computing," *IEEE Computer Magazine*, August 2001.

[3]   Rappaport, T. S., J. H. Reed, and B. D. Woerner, "Position Location Using Wireless Communications on Highways of the Future," *IEEE Communications Magazine*, October 1996.

[4]   Kayton, M., and W. R. Fried, *Avionics Navigation Systems*, New York: Wiley-Interscience, 1997.

[5]   Kupper, A., *Location-Based Services: Fundamentals and Operation*, New York: Wiley, 2005.

[6]   Sayed, A. H., and N. R. Yousef, "Wireless Location," in *Wiley Encyclopedia of Telecommunications*, J. Proakis, (ed.), New York: John Wiley and Sons, 2003.

[7]   Mech, L. D., and S. M. Barber, *A Critique of Wildlife Radio-Tracking and Its Use in National Parks,* A Report to the U.S. National Park Service, February 6, 2002.

[8]   Krizman, K. J., T. E. Biedka, and T. S. Rappaport, "Wireless Position Location: Fundamentals, Implementation Strategies, and Sources of Error," *Proc. IEEE 47th Vehicular Technology Conference*, Phoenix, AZ, May 4–7, 1997.

CHAPTER 2
Basic Principles and Applications

In Chapter 1, Section 1.3, we referred to radar, the classic device for measuring distance and position of a target by wireless means. Here we are concerned with how the measurement accuracy is affected by basic system characteristics of bandwidth, noise, and measurement clock rate. A radar terminal transmitter sends a short pulse at a precisely known time. The pulse is reflected back and the terminal receiver measures time of arrival. The total distance traveled by the transmitted signal is the difference between the transmitted and received times multiplied by the propagation speed of the signal, which is the speed of light. The distance between the terminal and the target is one half of the measured distance, which was based on the total two-way time of flight.

The radar provides location expressed as two- or three-dimension coordinates of the target with respect to the terminal. It uses a continuously rotating directional antenna whose bearing and elevation at the instant of receiving the reflected wave is recorded. The relative location of the target is specified in degrees of elevation angle and azimuth and the measured one-way distance.

In this book the targets do not passively reflect signals impinging upon them. A target is a terminal in a two-way or networked radio communication system whose principal function is exchange of data. However, from the point of view of distance and location, as in radar, operation must provide a specified accuracy within a given electromagnetic environment and with appropriate equipment and radio channel characteristics of bandwidth, noise, and measuring clock rate. The accuracy with which the radar device can measure the reflected pulse's time of return or time of arrival depends on the resolution and precision of its measuring clock, and the channel bandwidth and noise. Azimuth and elevation angle accuracy are a function of antenna beamwidth and noise.

Most often, distance measurement system design involves compromises among parameters of accuracy, bandwidth, clock rate, measurement time, and complexity. High accuracy in a short time needs a high clock rate and consequently high bandwidth. A large bandwidth, in turn, means greater noise power and reduced range and high clock rates increase complexity, current consumption, and cost.

This chapter describes the effects of clock rate, channel bandwidth, and noise using a simplified model of a radar system and baseband pulses. Later chapters show ways of manipulating the trade-offs to get the desired resolution and accuracy in the results for several common distance measurement and location technologies. The radar example is not the only one relevant to distance measurement. Instead of comparing transmission time with received time of arrival of a pulse, the phase of transmitted and received signals may be compared to determine time of flight.

In this case, narrow bandwidth is required for accuracy, as opposed to wide bandwidth for accurate measurement of pulse arrival time. Relative signal strength is also a criterion for estimating distance in some systems, and the significant environmental and system characteristics that affect performance differ from those applicable to distance measurement or location systems based on time-of-flight methods. Phase comparison and received signal strength methods are dealt with in Chapters 5 and 6.

The workings of distance measurement and location technologies can be illustrated by several legacy navigation systems. Most of these are being replaced by the global positioning system (GPS), but an understanding of their principles is very helpful for designing systems for areas where GPS is not a suitable platform. A description of the most important of these systems, as well as GPS, is given at the end of this chapter.

2.1 Signal Parameters

The ability to use a radio transmission for distance measurement and location and the accuracy that can be achieved depend on basic parameters of the signal, as well as the nature of its propagation.

2.1.1 Time Resolution

Let's return to the radar example of the first chapter. Figure 2.1 is similar to Figure 1.1 except that the initiator's time base clock has been added. A pulse is transmitted

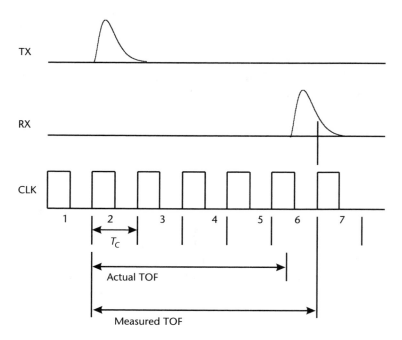

Figure 2.1 Time resolution.

and the time interval until reception is measured. The distance resolution is directly proportional to the period of the clock. The time of arrival (TOA) of the return pulse in the receiver can be distinguished only on a rise time of a clock pulse, so the time, and consequently the distance, is overestimated. The fall time could be used, but isn't because it would make the TOA dependent on another parameter—pulse duration.

Example: What is the one-way distance resolution, Δd, of a pulse radar system with a 10-Mbps clock?

(a) $T_c = 1/10$ MHz $= 100$ ns
(b) $c = 3 \cdot 10^8$ m/s
(c) $\Delta d = (T_c \cdot c)/2 = 15$m

The result, Δd, is obtained after dividing the total forward and back time-of-flight resolution by two to get the one-way resolution.

The minimum measurement period is equal to the time of flight. A short measurement time is essential when velocity must be determined and in general when the objects are in motion. We will see later that accuracy can be increased greatly over that determined from clock rate but at the expense of measurement time.

2.1.2   Pulse Width and Duty Cycle

There are two conflicting characteristics related to the pulse width. With a given peak power, a longer pulse means higher energy and higher S/N. However, the width of the pulse must be less than the round-trip flight time. Otherwise the initiator will still be transmitting when the echo returns, and the echo will not be received.

When pulse width and peak power are constant, reducing the duty cycle, or equivalently reducing the pulse repetition rate, lowers the average transmitted power, which may be a necessity to meet telecommunication regulations. When a low duty cycle is used in a network, multiple communication links can exist simultaneously on the same channel with a small probability of collision.

2.1.3   Bandwidth

The system bandwidth is a factor in the resolution of the time of detection. Note that while time resolution is involved, the influence of the bandwidth is different from that of the clock rate dealt with in Section 2.1.1 and therefore bandwidth is discussed separately here. The bandwidth referred to is the total bandwidth of the signal path between the generation of the pulse in the transmitter and its detection in the receiver. It includes, therefore, transmitter and receiver intentional and unintentional filtering, as well as the frequency response of transmitter and receiver antennas and that of the propagation path, which is not a constant function of frequency.

In effect the pulse rise time depends on the bandwidth, according to the relationship

$$B_{bb} = k \cdot \frac{1}{2 \cdot T_r}$$

where B_{bb} is the total noise bandwidth referred to baseband (one-half of the bandwidth in the RF passband) and T_r is the rise time. k depends on the particular transfer function that determines B_{bb} and on how T_r is defined. As a useful approximation we assume $k = 1$:

$$B_{bb} \approx \frac{1}{2 \cdot T_r} \qquad\qquad (2.1)$$

$$B_{bp} \approx \frac{1}{T_r}$$

where B_{bp} is the RF bandpass bandwidth, equal to twice the baseband bandwidth. The rise time is important because it creates an uncertainty as to the instant of arrival in a receiver of a transmitted pulse.

An important consequence of the bandwidth is its effect on multipath resolution. Radio signals arrive at a receiver over multiple paths because they are reflected from objects situated between the transmitter and the receiver. The paths of the reflected signals are longer than that of the direct, line of sight path, to a degree that depends on the distance of the reflecting object from the direct path. Accurate distance measurement depends on identifying the earliest arriving pulse, since its time of arrival is needed to find the true distance between transmitter and receiver. When bandwidth is relatively low, the rise time is long and the extended leading edge of the line of sight signal may be interfered with or smeared by pulses arriving along the multipaths thereby making it difficult to distinguish.

The bandwidth needed in a multipath environment depends on required accuracy, on the differences in path lengths of reflected echoes, and on the strength of the echoes relative to the line of sight signal. Assume a one-way distance accuracy of 3m is specified. A two-way time-of-flight resolution of $2 \cdot (3\text{m}/3 \cdot 10^8 \text{ m/s}) = 20$ ns must be achieved. A rule of thumb bandwidth is $B_{bp} = 1/20$ ns $= 50$ MHz. Bandwidth can be traded for measurement time in order to obtain a given accuracy with lower bandwidth than indicated by (2.1), when the interference is random noise. However, such a trade-off is not effective under multipath conditions, where path length differences are on the order of the required distance accuracy. Indoor systems are particularly affected by multipath interference because path differences can be several meters, equivalent to time of flight on the order of 10 ns, which is the accuracy that is frequently required. The influence of bandwidth on measurement accuracy in the presence of multipath is demonstrated in Figures 2.2 and 2.3. In both figures, there are three reflections delayed by 20, 40, and 60 ns from the direct path signal whose true time of flight is 40 ns. The line of sight amplitude is reduced by shadowing; that is, by relatively large objects in the direct path. The receiver in Figure 2.2 has an RF bandwidth of 50 MHz and that of Figure 2.3 a bandwidth of 20 MHz. In Figure 2.2 the time of arrival of the direct path signal is easily discerned, as it is the earliest pulse to be received. The line of sight (direct path) pulse peak is clearly seen at 50 ns, delayed 10 ns from the time of arrival of

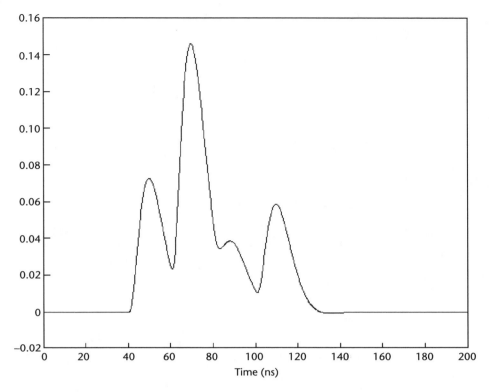

Figure 2.2   Received multipath signals with a bandwidth of 50 MHz.

the beginning of the leading edge. However, with reduced bandwidth of 20 MHz, shown in Figure 2.3, the direct path signal peak cannot be recognized and the estimate of the time of arrival of the pulse will be too high because of the influence of the later arriving multipath signals.

2.1.4   Noise

Noise is the ultimate limitation of communication efficiency and range. It also is the limiting factor in determining the time of arrival of a pulse. A small bandwidth reduces noise while a large bandwidth increases it.

Figure 2.4 will help explain the relationships between time-of-arrival estimation accuracy, signal-to-noise ratio, bandwidth, pulse width, and noise. The pulse in the figure is one of a train of pulses reflected or retransmitted from a target whose distance from the receiver is to be estimated. Figure 2.4 shows the baseband pulse after demodulation. Assuming that the pulse originated as a square wave at the transmitter, the rise time T_R is a function of the receiver bandpass bandwidth, B_{bp}:

$$T_r = 1/B_{bp} \qquad (2.2)$$

In the absence of noise, pulse arrival time t_0 can be measured exactly if the amplitude and threshold remain constant. There is a delay due to pulse rise time that is constant from pulse to pulse, so it can be subtracted from t_0 when calculating the

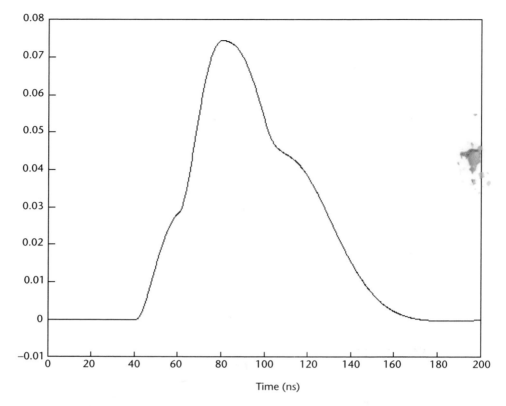

Time (ns)

Figure 2.3   Received multipath signals with a bandwidth of 20 MHz.

actual time of flight. Noise, however, causes the measured time to differ from t_0 by Δt_R, which is a random variable. If the signal-to-noise ratio (S/N) is large, we assume that the slope of the leading edge of the curve of the pulse plus noise is the same as that of the noiseless pulse [1]. From Figure 2.4:

$$\frac{n}{\Delta T_R} = \frac{A}{T_R} \tag{2.3}$$

Let σ_t be the rms error in the measurement of t_0:

$$\sigma_t^2 = \overline{\Delta T_R^2} = \overline{\left(\frac{n}{A}\right)^2} \cdot T_R^2 \tag{2.4}$$

where $\overline{x} = E(x)$ is the statistical average, or expectation. The average noise is zero, that is, $\overline{n} = 0$.

The signal-to-noise ratio relating to a sinusoidal carrier is

$$\frac{S}{N} = \frac{1}{2} \cdot \frac{A^2}{\overline{n^2}} \tag{2.5}$$

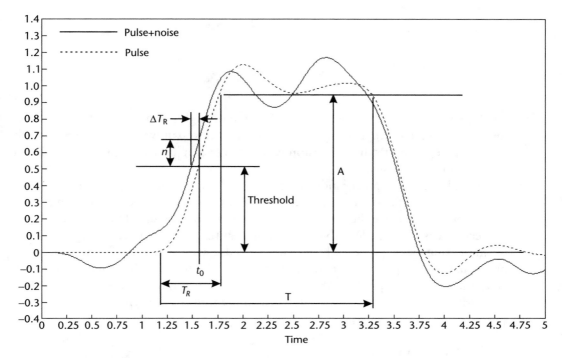

Figure 2.4 Filtered pulse and noise.

which when substituted in (2.4) and taking the square root gives the standard deviation of the timing error:

$$\sigma_t = \frac{T_R}{\sqrt{2 \cdot \dfrac{S}{N}}} \qquad (2.6)$$

It is often useful when discussing the parameters of pulse transmissions to refer to the pulse energy to noise density ratio E/N_0 instead of carrier signal power to noise power ratio. The noise power is $N = N_0 B_{bp}$ and the energy in the rectangular pulse is the signal power S times the pulse width T (shown in Figure 2.4): $E = ST$. Substituting these relations, as well as (2.2), in (2.6), we get:

$$\sigma_t = \frac{1}{\sqrt{\dfrac{B_{bp}}{T} \cdot \dfrac{2E}{N_0}}} \qquad (2.7)$$

Note that the coefficient $(B_{bp}/T)^{1/2}$ has the dimensions of bandwidth. Skolnik [1] calls this an effective bandwidth, or rms bandwidth β. Then

$$\sigma_t = \frac{1}{\beta \cdot \sqrt{2 \cdot \dfrac{E}{N_0}}} \qquad (2.8)$$

where

$$\beta = \sqrt{\frac{B_{bp}}{T}} = \sqrt{\frac{1}{T_R \cdot T}} \qquad (2.9)$$

Equation (2.8) shows clearly that the time-of-flight accuracy, and correspondingly the estimated range, is inversely proportional to bandwidth. Wide bandwidth systems give the best range precision. Common examples are spread spectrum and ultrawideband systems.

The above development indicates that while keeping the energy-to-noise density ratio constant, a narrow high-power pulse gives better range accuracy than a wide lower-power one since the narrow pulse has wider bandwidth. However, the high-power pulse is often undesirable. Assuming that the average power is kept constant by increasing peak power while reducing pulse width, the system with narrow pulses has a high peak to average power ratio, which is disadvantageous in many systems, since it means that the transmitter power amplifier must accommodate the high-power pulses even though the average power is relatively low.

2.1.5   Pulse Compression

The method of estimating the pulse arrival time by noting the instant that the input signal crosses a given level is prone to frequent false alarms, since noise bursts or low energy extraneous signals could have a brief amplitude that exceeds the threshold. In order to achieve maximum probability of signal detection and minimum probability of false alarms for a given energy to noise density ratio, E/N_0, the detector should consist of a matched filter. In the case of a rectangular pulse, this filter is an integrator with a discharge switch that resets the integrator at the end of the pulse duration. Since the pulse arrival time is initially not known, the receiver searches for pulses in a pulse train by adjusting the instant of the start of integration until the output of the integrator reaches a peak value just before the closing of the switch. In this case, the minimum passband bandwidth is $2/T$, where T is the pulse duration. Figure 2.5 compares the arrival time resolution of two pulses having the same energy. Both pulses give the same output of the matched filter if timing is precise. However, the short, high-amplitude pulse in Figure 2.5(b) gives better resolution because the slope of the integrated signal is greater than that of the pulse in Figure 2.5(a). The bandwidth needed to pass the signal in Figure 2.5(b) is four times greater than that needed for the signal in Figure 2.5(a), which is consistent with (2.8).

We see that the short pulse system has the higher range precision, even though both the short and long pulse systems have the same pulse energy. Average power per pulse is maintained by increasing peak power of the short duration pulse. The pulse repetition rate is the same in both cases.

It is possible to maintain a reasonable peak to average power ratio using a wide pulse, while increasing the bandwidth considerably in order to get good time-of-arrival precision. The method of doing this is called pulse compression. Two common pulse compression methods used for ranging are chirp modulation and

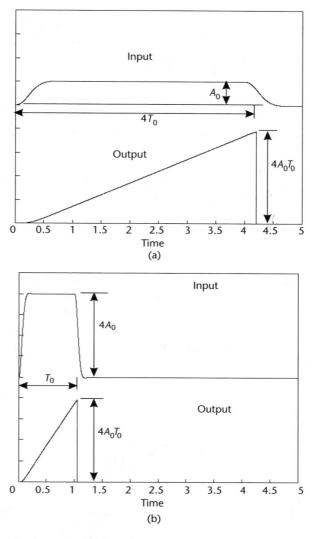

Figure 2.5   Received pulses at the input and output of an integrator matched filter. (a) Long pulse, $T = T_0$; and (b) short pulse, $T = T_0/4$.

direct sequence spread spectrum. In addition to giving increased range precision, they both discriminate against interference.

A chirp pulse is created by frequency modulating the pulse with a linearly changing (increasing or decreasing) sawtooth baseband signal, expressed as follows:

$$s(t) = \sin\left[2\pi \cdot \left(f_0 + \frac{k}{2} \cdot t\right) \cdot t + \phi\right] \quad 0 \le t \le T \tag{2.10}$$

where $s(t)$ is the chirp pulse, f_0 is the start frequency, k is the rate of frequency change per unit time, φ is a random phase, and T is the pulse width. A chirp pulse with $k = 20$ is shown in Figure 2.6.

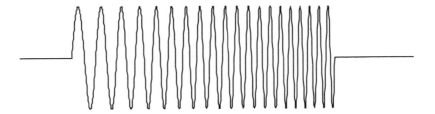

Figure 2.6 Chirp pulse. Rate of frequency change $k = 20$.

The chirp signal is detected using a matched filter. The impulse response $h(t)$ of a matched filter is a delayed and reversed version of the input pulse, expressed as

$$h(t) = s(T - t) \tag{2.11}$$

Figure 2.7 shows the single pulse spectrum and detector output of signals with the same pulse width T and start frequency f_0, but different values of chirp parameter k. Detector outputs are the squared outputs of matched filters. All signals have the same energy but the time resolution of the detector outputs is inversely proportional to the bandwidth, which is a function of the parameter k. The signal of Figure 2.7(a) has the widest spectrum and sharpest pulse arrival time resolution.

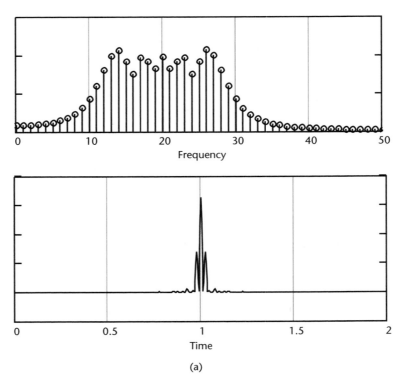

Figure 2.7 Chirp pulse spectrum and time resolution for different values of chirp parameter k. (a) $k = 20$, (b) $k = 5$, and (c) $k = 0$.

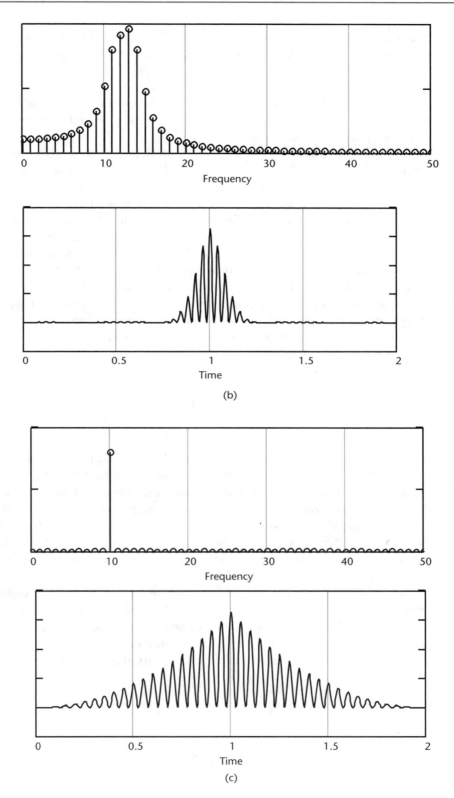

Figure 2.7 (continued).

In Figure 2.7(b), the bandwidth is approximately one-fourth as large as that of Figure 2.7(a), and time resolution is around four times worse. A constant frequency pulse shown in Figure 2.7(c) has the same energy as the two chirp pulses but does not provide their advantages of time resolution and interference rejection. It is evident that in order to use the improved time resolution obtainable from the wide bandwidth chirp signals, the receiver clock rate must be high enough to detect the compressed matched filter output pulse.

The matched filter for chirp pulse generation and detection is commonly implemented using a SAW dispersive delay line that is fabricated specifically to match the known parameters of the signal. A dispersive delay line has a propagation time between input and output that is a function of signal frequency.

A second method for increasing bandwidth while maintaining a constant pulse duration at a given pulse energy is based on direct sequence spread spectrum. In the transmitter, the RF pulse carrier is modulated by a sequence of bits that have very good autocorrelation properties. These bits, which are used for pulse compression and not directly as data, are called chips. The received signal, $r(t)$, which contains added noise and interference, is cross correlated with a locally generated sequence $s(t)$ that corresponds to the expected chip sequence. The correlation process over a pulse duration T can be expressed as

$$Z(T) = \int_0^T r(t)\,s(t)\,dt \tag{2.12}$$

When $r(t)$ and $s(t)$ are similar over a period of T and are lined up in time phase, $Z(t)$ will have a maximum value that is proportional to the energy of the received signal over the pulse width T. If $s(t)$ has good autocorrelation properties, the output $Z(T)$ will be relatively small compared to its maximum value when $r(t)$ is shifted in time by one chip or more relative to the locally generated $s(t)$. The output of the correlator, then, is a compressed pulse with average width of one chip that has the same equivalent energy as the input pulse of width T. Time-of-arrival resolution is ±1 chip.

An example of good pulse compression sequences are Barker codes, listed in Table 2.1 for $N = 5, 7, 11$, and 13, where N is the number of bits in a sequence. The bits are bipolar and therefore are shown as sequences of plus and minus symbols. Note that the bits in a sequence may be inverted, or the sequence may be reversed, without affecting the cross correlation properties.

Table 2.1 Barker Codes for
$N = 5, 7, 11, 13$

N	Sequence
5	+ + + − +
7	+ + + − − + −
11	+ + + − − − + − − + −
13	+ + + + + − − + + − + − +

Equation (2.13) is a discrete expression for (2.12) where k is the shift in chips between the two sequences, s_i is a chip of the local generated sequence, and r_{i+k} is a received chip at the sample time i.

$$Z_k = \sum_{i=1}^{N} r_{i+k} s_i \qquad (2.13)$$

Table 2.2 shows the values of Z_k for $k = 0 \ldots 6$ using the 7-bit Barker code of Table 2.1, calculated with (2.13). The Z_ks when k equals 1 to $N - 1$ are called sidelobes. The first row is the prototype sequence $\{s_i\}$ and the following rows are the shifted chip sequences $\{r_{i+k}\}$, free of noise and interference. The sidelobes vary between −1 and 0, and the correlation when the sequences line up at $k = 0$ is 7. This shows that the input pulse with Barker code modulation whose energy is spread over seven chips has a seven times improvement in time-of-arrival resolution at the output of the correlation process. Similar improvement, in proportion to the value of N, is obtained with other Barker sequences and different codes with good autocorrelation properties.

In the simplified spread spectrum receiver block diagram Figure 2.8, a baseband matched filter implements the correlator. It is followed by a sliding window low-pass filter. When the expected spread spectrum pulse is received, the digitally filtered output of the matched filter exceeds the threshold of the detector, which

Table 2.2   Correlation Values for 7-Chip Barker Code

$s_1 = 1$	$s_2 = 1$	$s_3 = 1$	$s_4 = -1$	$s_5 = -1$	$s_6 = 1$	$s_7 = -1$	k	Z_k
−1	0	0	0	0	0	0	6	−1
1	−1	0	0	0	0	0	5	0
−1	1	−1	0	0	0	0	4	−1
−1	−1	1	−1	0	0	0	3	0
1	−1	−1	1	−1	0	0	2	−1
1	1	−1	−1	1	−1	0	1	0
1	1	1	−1	−1	1	−1	0	7

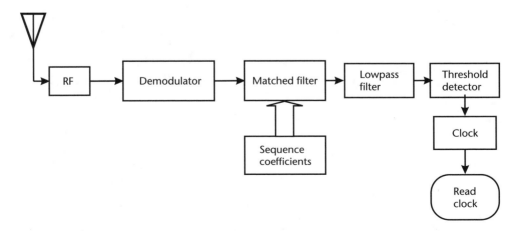

Figure 2.8   Spread spectrum pulse receiver with matched filter.

outputs the value of the real time clock, which is an estimate of the time of arrival of the pulse.

Figure 2.9 is a digital correlator based on the 7-chip Barker code. Note the direction of the input chips, and that the locally stored sequence is in reverse order to that direction. This is in conformance with the impulse response of the matched filter, given in (2.11). The matched filter consists of six 1-bit delay elements, which could be implemented by shift registers, multipliers and an accumulator.

A simulation output of the spread spectrum pulse arrival time estimator with a 7-chip Barker code is shown in Figure 2.10. Noise was added to the input signal

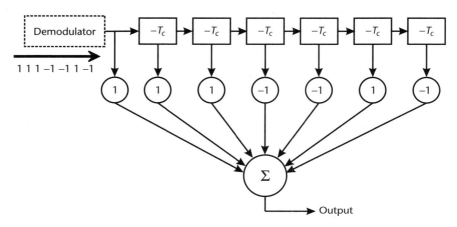

Figure 2.9   A 7-chip digital matched filter.

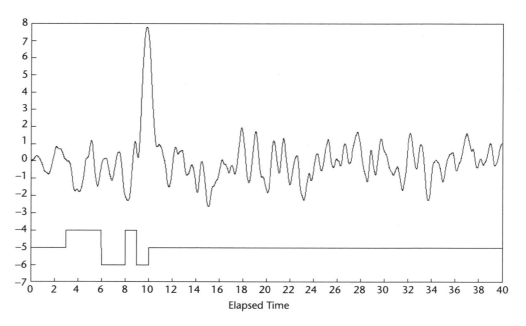

Figure 2.10   Result of simulation of arrival time detection by matched filter of a 7-chip spread spectrum sequence in noise.

for $E/N_0 = 18.5$ dB. Note that the matched filter output peak is clearly distinguished from sidelobes and noise.

Frame synchronization in burst type data communication systems is often achieved using the spread spectrum correlation technique described in the preceding paragraphs. An 11-bit Barker code is the basis for despreading the data in 1 and 2-Mbps IEEE 802.11 DSSS physical layer used in WLAN. Other protocols include a frame synchronization sequence as part of the packet preamble, or of every data frame. Usually these sequences are more than 13 chips long, for which there is no Barker code, so other sequences with good correlation properties are employed. The frame delineation epoch is a convenient place to make a time-of-arrival estimation for TOA and TDOA distance measuring and location methods.

2.2   Basics of Location

Four geometric arrangements for calculating location coordinates are described next [2]. They represent the different ways of finding location from combinations of the basic measurements of distance, represented by the Greek letter rho, and the angle of arrival, theta.

2.2.1   Rho-Theta

When both direction finding (DF) and distance measurement capability is available, only one fixed terminal is needed to determine the position coordinates of the target as shown in Figure 2.11. The target is located on the intersection between a circle whose radius is ρ, the distance between fixed terminal F and target T, and a bearing line that is at an angle of θ referenced to north. The directional antenna, as shown in the diagram, is located at the fixed terminal, but the direction finding capability may be located at the target, as in the case of the very high frequency

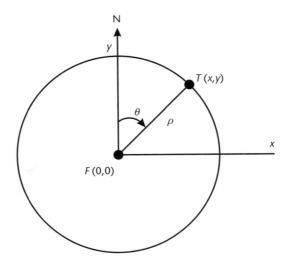

Figure 2.11   Rho-theta location measurement configuration.

omnidirectional ranging (VOR) navigation aid (see Section 2.3.2). If the fixed station estimates distance by receiver signal strength, the perimeter line may not be a circle but a constant signal strength contour based on a mapping of path loss in the region of the terminal. If F is the origin, the coordinates of T are

$$x = \rho \cdot \sin(\theta)$$ (2.14)
$$y = \rho \cdot \cos(\theta)$$

Examples of distance-angle location are VOR combined with distance measuring equipment (DME) (see Sections 2.3.1 and 2.3.2), wildlife location where distance is approximated by signal strength (Section 8.4.2.1 and [3]), and some systems of article location in a warehouse.

2.2.2 Theta-Theta or AOA

Directional antennas can be used at two or more fixed terminals to find target location when the coordinates of the terminals are known relative to a reference point. The geometric procedure for calculation location is called triangulation. An advantage of this method is that target direction can be found without any time synchronization or restrictions of modulation type or protocol of the transmitted signals. Figure 2.12 shows a mobile transmitting target T and two fixed stations F_1 and F_2 with directional antennas. The coordinates of F_1 and F_2 are known and the angles of arrival, θ_1 and θ_2 of the signal referenced clockwise from north are measured. With the origin at F_1 the coordinates of T are:

$$y = \frac{y_2 \cdot \tan(\theta_2) - x_2}{\tan(\theta_2) - \tan(\theta_1)}$$ (2.15)
$$x = y \cdot \tan(\theta_1)$$

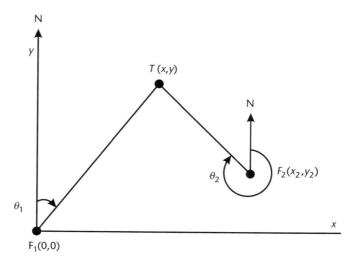

Figure 2.12 Theta-theta location measurement configuration.

The accuracy of the location position depends primarily on the directivity of the antennas. Generally, directive antennas are significantly larger than omnidirectional antennas. For automated location, electronically steered antennas are used for direction finding. Theta-theta is often used for locating wild animals where a wireless distance measurement is not available.

2.2.3   Rho-Rho or TOA

In both the rho-rho/time-of-arrival (TOA) and time-difference-of-arrival (TDOA) arrangements, directional antennas are not used and location is found by trilateration using distance data only. Distance can be estimated using received signal strength (RSS) data, or time-of-flight measurements. Assume the transmitter and receiver have synchronized clocks. Then the receiver can find the time of flight by subtracting the time the signal is received from the time of transmission. At least two fixed terminals are needed to locate a target in two dimensions. Three or more fixed terminals must be available for three-dimensional location. In a unilateral arrangement, the fixed stations are beacon transmitters, and the target is a receiver. Multilateral systems use fixed terminal receivers to estimate the distance to a transmitting target.

The geometry of determining two-dimensional location from distance measurements is shown in Figure 2.13. The coordinates of two fixed terminals, F_1 and F_2 are known in a given frame of reference specified by the x-y axis and the origin at F_1. If we can find the distances ρ_1 and ρ_2, we can determine the coordinates of T from a point of intersection of two circles. Since the circles intersect in two locations, we will assume the ambiguity is solved by knowing that T is in the upper half of the x-y plane. If there is no other knowledge to eliminate the ambiguity, a third fixed terminal is required.

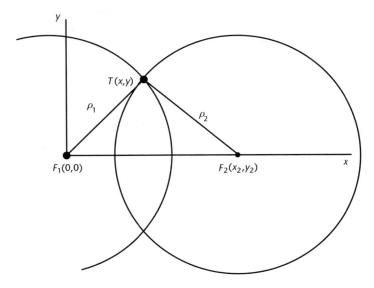

Figure 2.13   Rho-rho location measurement configuration.

Using the method denoted by time of arrival (TOA) or RSS, we find the distances ρ_1 and ρ_2. In the case of TOA, the one-way distance between T to F_1 or F_2 is determined as follows. Assume that all three stations have high precision clocks that are set to exactly the same time and that F_1, F_2 are receivers. A pulse sent from T at time t_0 is received at F_1 at time t_1 and at F_2 at time t_2. T notifies F_1 and F_2 of the time of transmission, t_0, by time-stamping its message. Now F_1 and F_2 can calculate the distances ρ_1 and ρ_2 from the transmit and receive times and the known propagation speed, the speed of light, c.

$$\rho_1 = (t_1 - t_0) \cdot c \qquad\qquad (2.16)$$
$$\rho_2 = (t_2 - t_0) \cdot c$$

The equations of the two circles are

$$\rho_1^2 = x^2 + y^2 \qquad\qquad (2.17)$$
$$\rho_2^2 = (x - x_2)^2 + (y - y_2)^2$$

These two nonlinear equations can be solved to find x, y.

2.2.4 TDOA and Hyperbolic Curves

While TOA gives a straightforward way to find location from distance measurements, it does have disadvantages for many applications. Accurate, synchronized clocks must be maintained in all stations participating in the measurements. Information must be passed from the initiator to the receiver specifying when the transmission was started. Another geometric location method, TDOA, does not have these disadvantages. All it needs is a transmission that has a recognizable unambiguous starting point. The data used in the location calculations is the time difference in the reception of that starting point at the several base stations, and not the actual time of flight of the signal from the target to the fixed stations. In an arrangement having a mobile target whose coordinates are to be determined and two fixed base stations, as we had in the example of TOA, we can find the time difference of arrival of a signal sent from the mobile and received at the base stations. This one time difference value is not enough to calculate the two coordinate values of the mobile's position. So, in order to have sufficient data to find two unknowns—the mobile's coordinates—TDOA requires one more base station than TOA. The clocks of the fixed stations must be synchronized, but not that of the target.

TDOA is used unilaterally—the target finds its own position from fixed station transmissions—or multilaterally, where time difference data is collected from target transmissions by fixed base station receivers. An example of the former is Loran-C, described in Section 2.3.3. Cellular network–based systems use multilateral TDOA.

Figure 2.14 shows the geometric layout for TDOA in two dimensions. Target T transmits a pulse at t_0 that is received at F_1 at t_1 and at F_2 at t_2. The clocks of F_1 and F_2 are synchronized, but T's clock is not, so t_0 is not known. However, the time difference of arrival, which is $t_2 - t_1 = (t_2 - t_0) - (t_1 - t_0)$, can be

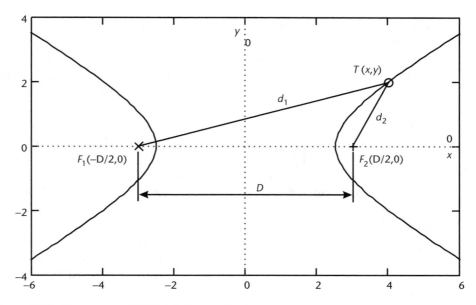

Figure 2.14   Geometry of TDOA location method.

calculated. The times on the right side of the equation are proportional to the distances d_1 and d_2 shown in Figure 2.14 since the distance is the time of flight times the speed of light, c. Therefore the difference of the distances between the two fixed stations and the target is $\Delta d = d_2 - d_1 = c(t_2 - t_1)$. When all stations are in a plane, the locus of points whose difference of distances from F_1 and F_2, Δd, is constant describes a hyperbola. Thus, the time difference of arrival that is obtained from times of arrival measured at two synchronized fixed stations indicates that the target is located somewhere on a hyperbola. The particular branch of the hyperbola that the target is on is the one that is closest to the fixed station that received the signal first. Figure 2.14 is drawn with F_1 and F_2 on the x-axis and each at equal distance, $D/2$, from the origin. The expression for the hyperbola is

$$\frac{x^2}{a^2} - \frac{y^2}{b^2} = 1 \tag{2.18}$$

Expressing a and b in terms of the known quantities Δd and D, we have

$$a^2 = (\Delta d/2)^2 \tag{2.19}$$

$$b^2 = \left(\frac{D}{2}\right)^2 - a^2 \tag{2.20}$$

The generality of these expressions is not affected by the convenient way we drew the deployment of the stations because any rectangular coordinate system can be converted by formulas of translation and rotation [4].

Since the time difference of arrival found from TOA measurements by two terminals places the target on a locus of positions, it is necessary to use the time of arrival at a third fixed station, t_3, to pinpoint the target location. With the addition of this one station, we can now find three time differences of arrival: between F_1 and F_2, F_2 and F_3, and F_1 and F_3. The intersection of a minimum of two hyperbolas, constructed from two times of arrival determinations and drawn on the same coordinate system, gives the location of T, as shown in Figure 2.15. The second hyperbola, shown as a solid curve, is based on the time difference of arrival between F_2 and F_3.

The example in the above development was assumed to be a multilateral system with a target transmitter and fixed terminal receivers. The geometrical determination of the target location would be exactly the same for a unilateral system where the target measures times of arrival of signals from three or more fixed stations whose locations are known. However, the transmission times of the fixed stations must be staggered so that their transmissions do not interfere with one another. When the target knows the transmission time of each station, it can estimate the required time differences of arrival.

2.3 Navigation Systems

In order to demonstrate some ways the various methods of distance measurement and location are used, we examine some navigational applications. Some of these were put in use around the time of World War II but are still used today, although they are being phased out by GPS.

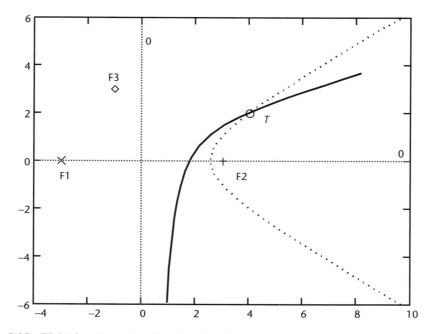

Figure 2.15 TDOA location using three fixed stations.

An excellent demonstration of the combination of the principles of time of flight and angle of arrival to find range and position is provided by two classical navigation methods in use since World War II—DME and VOR [2]. The ground beacons of these two methods are usually located at the same site. DME measures range between an aircraft and the site, and VOR provides an angle bearing, referenced to the magnetic north pole.

2.3.1   DME

The use of the DME is illustrated in Figure 2.16. Its operation is similar to secondary radar, where a transponder in the target retransmits the incoming signal back to the ground terminal. In the case of DME, an interrogator in an aircraft sends repeating pairs of pulses to a transponder navigation beacon at the ground based station. The transponder sends back reconstituted pulses to the interrogator. The interrogator measures the net elapsed time between its transmission of the pulses and their reception, computes the path distance (slant range) and displays this distance on a cockpit display. Accuracy is better than the greater of ±0.5 nautical miles or 3% of range in 95% of measurements [5]. Also displayed is velocity relative to the beacon, and estimated aircraft time to the beacon. The operational details are as follows.

DME uses frequencies over a band from 978 to 1,213 MHz, contained in two sets of 126 channels, each spaced 1 MHz apart. Transmit and receive channels for both the aircraft interrogator and DME beacon transponder are separated by 63 MHz. The DME unit automatically is set to a specified transmit-receive channel pair when the aircraft pilot selects a VOR channel in the 108 to 117.95 MHz band. The interrogator transmits pairs of pulses repeated at a rate of around 30 pairs of pulses per second. Pulse pair timing is shown in Figure 2.17. Individual pulses are 3.5 μs wide and the separation of the pulses of a pair, T_1, is 12 μs or 36 μs apart, depending on the particular beacon specification (X mode or Y mode).

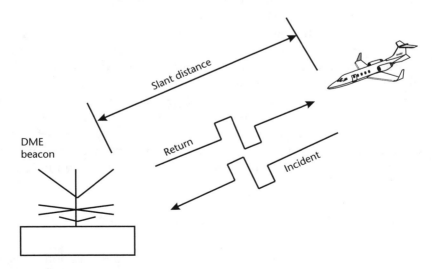

Figure 2.16   DME ranging.

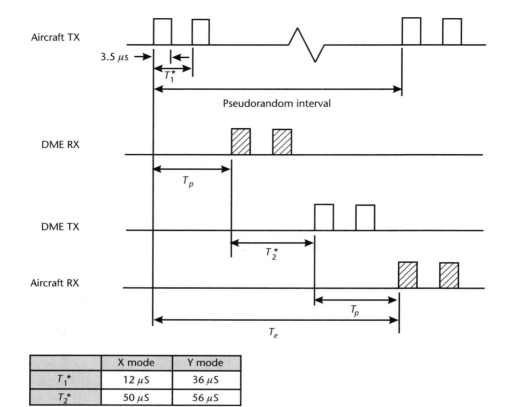

	X mode	Y mode
T_1^*	12 μS	36 μS
T_2^*	50 μS	56 μS

Figure 2.17 DME pulse timing.

The use of pulse pairs with known time between the individual pulses allows the transponder and interrogator receivers to ignore random pulses that are not part of the system. In Figure 2.17, T_p is the one-way propagation delay and T_e is the total elapsed time at the interrogator between pulse pair transmission and reception of the echo from the DME transponder.

When the DME beacon receives a pair of pulses, it retransmits them after a precise delay, T_2, of 50 μs (X mode) or 56 μs (Y mode). This delay is necessary in order to prevent a "reflected" first pulse of a pair from reaching the interrogator before the second pulse is transmitted, which could happen at short distances. In computing range, the interrogator measures the total elapsed time, T_e and subtracts the transponder delay from the measured time between a transmitted pulse pair and its reception. The remainder time is the two-way time of flight, which when divided by two and multiplied by the speed of light gives the slant range between the aircraft and the beacon. The range expression is shown in (2.21), where c is the speed of light.

$$slant\_range = \frac{T_e - T_2}{2} \cdot c \qquad (2.21)$$

Several aircrafts could be interrogating a DME ground station transponder at the same time, so each interrogator must have a way to distinguish its own retransmitted reflections from other received pulse pairs. It does this by varying the interval between pulse pairs in a pseudorandom fashion whose pattern differs for every interrogator. The average pulse pair repetition rate remains approximately 30 pulse pairs per second. The way each interrogator recognizes only its own pulse reflections is shown in Figure 2.18. The interrogator uses a range gate to open a narrow window at the time when echoes from its transmissions are expected. The time interval from interrogator pulse transmission to the opening of the range gate window is varied during a search of transponder replies. During trial 1 of Figure 2.18, the window is open on a return belonging to a different aircraft interrogator. Signal energy is accumulated during subsequent windows that are spaced in time according to the interrogator's pseudorandom pattern. Echoes from other transponders will be received outside of the window most of the time. In the case of trial 1, the threshold is not reached and the range gate interval is adjusted to receive another pulse. In range gate trial 2, a correct return is found, and energy accumulates during several successive windows until the threshold is exceeded. At this point, the elapsed time since the previous transmission is measured and the slant range is calculated according to (2.21).

This process is actually a check of correlation between the range gate window pattern and the received echoes. When the range is correct, determined by the phase or position of the range gate window pulse train, the power in the range gate windows accumulated over a given number of pulses reaches a peak value and the range reading on the cockpit indicator can be updated. Note that the range gate window encompasses only one of the two pulses in the reflected pair.

The accuracy of the range estimate may be enhanced by using a range gate with two windows as shown in Figure 2.19. The windows are coupled so that their opening times correspond to the estimated range. The width of each window is two pulse widths, and their overlap is just shorter than the width of one pulse. The input signal is integrated over the time window of each gate and the two results are compared. If the range gate position, t_{RG}, is earlier than the actual

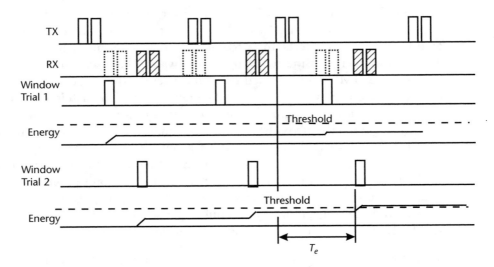

Figure 2.18   Range gate correlation of DME return echoes.

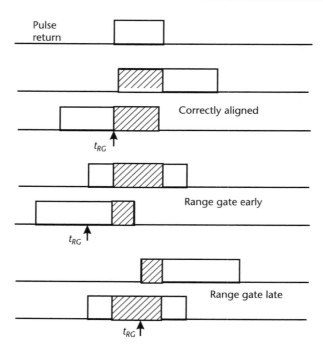

Figure 2.19 Dual window range gate.

pulse arrival time, the resulting energy in the far window will be greater than that in the near window, and if its position is too late, the near window energy will be greater. A feedback signal that is proportional to the difference of the measured energies in the two windows is applied to adjust the position of the range gate for the next pulse that is expected and the process is continued until each window has equal energy and the energy in the overlap is maximum. The reference position of the coupled windows, t_{RG}, at this time then gives the actual time of arrival of the pulse, from which the range can be determined.

The ultimate accuracy depends on the clock rate of the counter that measures the position of the range gates. In the case of the double range gate system, the signal-to-noise ratio affects the accuracy since it determines how accurate the differences of energy in the two range windows can be discerned.

When the DME is first turned on in the aircraft, the range is not known and the range gate position has to be moved incrementally after each interrogator pulse pair transmission until a return pulse is detected in a range gate window. In the case of a single range gate with window size equal to the pulse width, the window position increment should be one half of a pulse width, to be sure a pulse may always be detected. The worst-case acquisition time depends on maximum range and the pulse repetition rate. The DME may measure a slant range up to 130 nm at elevation above 18,000 feet. The repetition rate during acquisition is considerably higher than the average, typically 150 pulses per second. With one half pulse width equal to 1.75 μs we can calculate acquisition time as follows:

1. The longest two-way time of flight = 130 nm × 2 × 1,852 m/nm/3 × 10^8 m/s
   = 1.6 ms

2. Longest acquisition time = (1.6 ms/1.75 μs)/150 pps
                              = 6.1 seconds

This time can be reduced considerably by several means. A better algorithm for moving the range window could be used, for example by starting with maximum range and decrementing the window position, instead of increasing it from zero. A wider range gate window would decrease search time, while increasing the probability of a false pulse detection. Choosing window size according to received signal strength reduces the disadvantage. Conducting the echo search on several independent parallel acquisition channels, each with a different window opening time, will reduce acquisition time in proportion to the number of parallel channels. Once a target DME has been acquired, the aircraft instrument can maintain almost instantaneous tracking with changes of relative speed to the beacon, and can calculate this speed as well as expected time of arrival at the beacon. A modern DME can lock on to a beacon in less than one second, measuring distance up to 300 nm and ground speed to 999 knots.

2.3.2 VOR

The principle of VOR (VHF omnidirectional ranging) operation demonstrates how direction may be determined to a precision that doesn't depend directly on the beam width. It also provides an example of how phase comparison can be used in achieving that precision.

A VOR indicator in an aircraft helps the pilot maintain his bearing relative to the magnetic north pole. Angle of arrival radials are pointed away from the VOR, as shown in Figure 2.20. In the diagram, the plane is flying toward the beacon on the 135° radial emanating from the VOR station. The cabin display should indicate the plane's bearing as 135° + 180° = 315° with a flag sign TO. If the plane was in the same position but flying in the opposite direction, the indicator would still show 315°, flag TO, even though its heading is away from the VOR. In order for the display to indicate correctly, the OBS (omni-bearing selector) knob must be

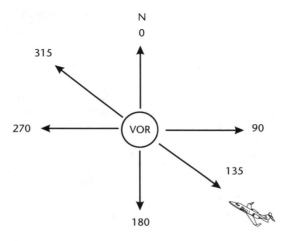

Figure 2.20 VOR bearing display.

set for the desired navigation track. The plane's bearing is the same whether the pilot is flying to the beacon or away from it, although the AOA radials detected by the aircraft VOR are 180° apart. The possible 180° ambiguity is eliminated by setting the instrument dial appropriately for the desired course.

The VOR operates on frequencies from 108 to 117.95 MHz in 50-kHz channels. Maximum range varies from around 25 to 130 nautical miles, depending on height and terrain. Overall accuracy is 4.5% [5]. The instrument came into widespread use in the 1950s and even today it is probably the most common navigational aid in operation, although it will be gradually phased out by GPS.

VOR uses a relative phase measurement between two 30-Hz signals demodulated from the beacon transmission to indicate bearing. The variable beacon signal is radiated from an antenna array whose azimuth pattern rotates at the rate of 30 Hz. The pattern takes the form of a geometric curve called a limaçon, shown in Figure 2.21(a) [6]. The equation of this curve is

$$r(\theta) = b + a \cdot \cos(\theta) \tag{2.22}$$

where in the form shown in Figure 2.21, $b > a$. The average of $r(\theta)$ is b and a is its amplitude around b (geometric details in [4, p. 44]). The rotation of the pattern at one time was done mechanically, but modern systems use electronic steering. At the receiver, the change in received signal strength due to pattern rotation is the equivalent of amplitude modulation by a base band sine wave of 30 Hz, as illustrated in the bottom part of Figure 2.21. The phase of the amplitude demodulated signal in the aircraft receiver is a linear function of its angular position around the VOR beacon. The beacon also provides a reference 30-Hz signal to which the phase of the variable signal can be compared. This reference signal is in phase with the variable signal, which is due to the antenna pattern rotation, when the pattern maximum is pointing to the magnetic north pole. The reference 30-Hz signal is frequency modulated on a subcarrier of 9,960 Hz, deviation ±480 Hz. The subcarrier is in turn amplitude modulated on the VHF band carrier. In addition a Morse code station identification signal with a tone frequency of 1,020 Hz is amplitude modulated to the carrier. A block diagram of the receiver is shown in Figure 2.22. Filters separate the baseband components of the amplitude demodulation. The variable and reference 30-Hz signals are input to a phase detector, whose output is submitted to the VOR display.

The use of VOR and DME together gives a pinpoint position, as a classic rho-theta application. Many aircraft have two or three VOR receivers. When used simultaneously, the intersection of radials from two or more VOR beacons gives the aircraft's location, in accordance with the theta-theta method.

2.3.3 Loran-C

Loran-C is a long-range navigation position finder that operates on a center frequency of 100 kHz [2, 7]. It went into service in the late 1950s and is presently being superceded by GPS, although it will probably continue to be maintained for several more years as a primary aid to small maritime and airborne users and as a backup for others. The system is a good illustration of the principles of a hyperbolic

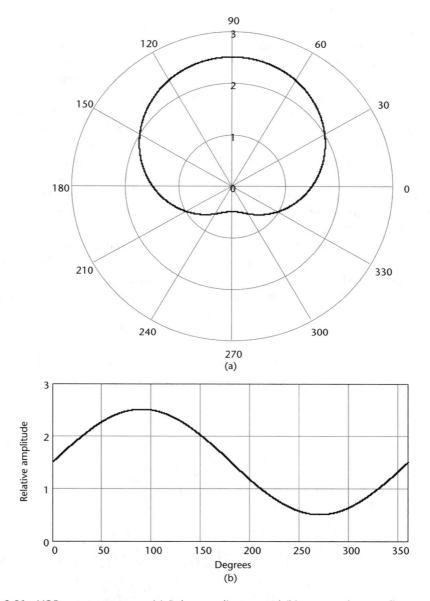

Figure 2.21   VOR antenna pattern. (a) Polar coordinates, and (b) rectangular coordinates.

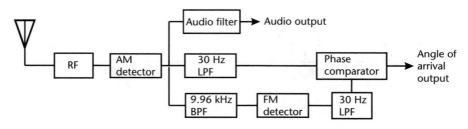

Figure 2.22   Block diagram of VOR receiver.

navigation aid. It is an example of a unilateral, one-way TDOA system, where signals from several transmitters are received at the point whose location is to be determined and where signal processing and position display are located.

Loran-C differs from many other time-of-flight systems by the fact that the instant of signal arrival is in most implementations determined directly from the received carrier time domain waveform, and not from a characteristic of a demodulated or down converted baseband signal. Loran-C signals are short pulses, and the time of arrival is measured at the zero crossing of one particular reference cycle of the RF wave. A Loran-C pulse is shown in Figure 2.23, where the reference point is the end of the third cycle. Since the carrier frequency is 100 kHz, zero crossings in a given direction, say, low to high, occur every 10 μs. Loran-C has a rated accuracy of 0.25 nm = 463m [7]. Choosing by mistake a crossover point adjacent to the reference one results in an error of 10 μs $\times$ 3 $\times$ 10^8 m/s = 3,000m.

As in all TDOA systems, Loran-C requires measurement of the difference in the propagation time over at least two pairs of paths between at least three fixed stations and the target. A Loran-C chain consists of one master station and two or more secondary stations deployed over distances of hundreds of kilometers between them. An example deployment is shown in Figure 2.24 [7] where the estimated coverage area is indicated by a dashed curve. The master station is referred to by the letter M and secondary stations by W, X, Y, and Z. A receiver that measures a time difference of arrival between a master and one secondary can

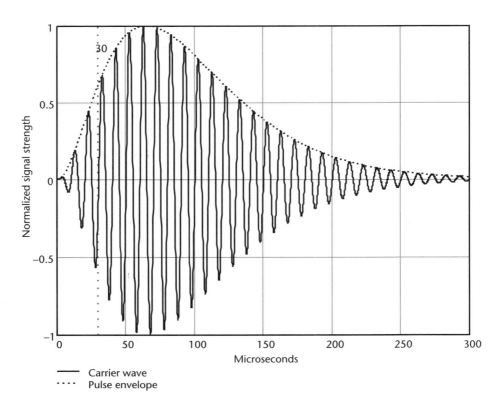

Figure 2.23   Loran-C pulse shape.

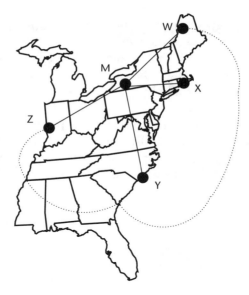

Figure 2.24 Northeast U.S. chain.

locate itself on a hyperbola whose foci are at the master and that secondary. Another time distance measurement between the same master and another secondary will place the receiver on another hyperbola. The intersection of the two hyperbolas is the location of the receiver. Loran-C is a two-dimensional system and considering its wide area coverage, the curvature of the Earth must be taken into consideration when presenting location coordinates in terms of longitude and latitude. A basic equipment displays the time differences of arrival of signals from a master and two secondary stations. The operator locates his or her approximate position at the crossing point of hyperbolas labeled with TDOAs from station pairs that are overlaid on geographic maps prepared specially for use with Loran-C.

The master and secondary stations transmit groups of pulses as shown in Figure 2.25. A master group is distinguished from a secondary group by containing nine pulses instead of eight. Pulses in a group are 1,000 ms apart. The last pulse in the master group is separated by 2,000 ms. A particular chain is identified by the time between successive master group transmissions, which is shown as T_{GRI} in Figure

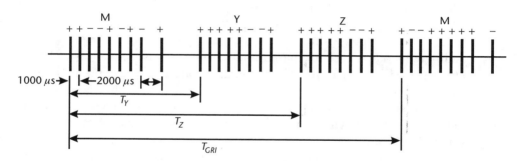

Figure 2.25 Loran-C pulse groups, timing, and phase codes.

2.25. The group repetition interval (GRI) is T_{GRI} in microseconds divided by 10. For example, the period of the master-secondary pulse sequence for the Northeast chain illustrated in Figure 2.24 equals 99,600 ms and the chain's GRI is 9,960. For each chain there are a set of time periods that define the delays of the secondary station transmissions from the master station's. In the case of the three-station chain of Figure 2.25, they are T_Y and T_Z. The Loran-C receiver at a target measures the elapsed time between the arrival of a pulse from the master station to the arrival of the corresponding pulse (among the eight in the group) from a secondary station. Letting τ_M be the propagation delay from master to target and τ_Y be the delay from secondary station Y, the measured time delay at station Y is

$$TD_Y = T_Y + \tau_Y - \tau_M \qquad (2.23)$$

The TDs for two secondary stations are located on the overlay maps mentioned above to get the target location.

In order to get the distance difference of arrival, Δd, needed to plot hyperbolas according to (2.18) to (2.20), the speed of propagation, v_P, must be known. Then

$$\Delta d = v_P \cdot (TD_Y - T_Y) \qquad (2.24)$$

Δd is positive if the target is closer to the master than to the secondary station; otherwise it is negative.

The speed of propagation is symbolized by v_P, and not c, the speed of light in a vacuum, because $v_P < c$ except in a vacuum and approximating it as c reduces accuracy. The true value of v_P depends on the topography between the Loran-C transmitters and the target. The groundwave, which is the principal mode of propagation at the 100-kHz carrier frequency, travels more slowly over water than through the atmosphere, and generally more slowly over land than over water. The Loran-C overlay maps are plotted taking the propagation time into consideration, and advanced receivers that directly display target coordinates can also account for the propagation time in their calculations.

Loran-C accuracy is based on groundwave propagation between the master and secondary stations and the target. However, signals are reflected from the ionosphere and could confuse the receiver as to the correct signal to use. This problem is alleviated by coding the phase of individual pulses in a pulse group. The sine wave carrier in each pulse of a group is sent shifted by 0° or 180° according to a code sequence. The coding is shown in Figure 2.25. The plus sign indicates that there is no phase shift, and the minus sign means that the signal is inverted. A different code is used for master and secondary pulse sequences and repeats itself every period of $2 \times T_{GDI}$. The secondary code in the second GRI period is not shown in the diagram. It is (+ − + − + + − −). When the received pulse group sequence is correlated by a locally generated sequence having the known code, delayed signals arriving via a skywave path, as well as interfering signals from other chains, are rejected. A receiver may distinguish a master pulse group by its unique code sequence, instead of by waiting to see if it has an additional, ninth pulse.

Estimating the time of arrival of a pulse at a reference epoch is crucial to the operation of Loran-C, as for any TOF method. The time of arrival instant is conventionally chosen to be the zero crossing at the end of the third period of the RF carrier. A magnified view of the Loran-C pulse is presented in Figure 2.26, where the reference crossover point at 30 μs from the start of the pulse is indicated by a vertical dotted line. The equation of the pulse envelope is

$$E(t) = A \cdot t^2 \cdot e^{-2\frac{t}{65}} \tag{2.25}$$

where t in microseconds is defined up to the envelope peak, $0 \le t \le 65$. The envelope for the fall time of the pulse may differ from (2.25), but Figures 2.23 and 2.26 show $E(t)$ expressed by (2.25) continuing beyond 65 ms. A is an amplitude constant whose value is set in the figures such that the pulse maximum is unity. $E(t)$ modulates a sine wave carrier of frequency $f_0 = 0.1$ MHz, so the transmitted pulse is:

$$x_0(t) = E(t) \cdot \sin(2 \cdot \pi \cdot f_0 \cdot t) \tag{2.26}$$

Finding this crossover by counting from the beginning of the signal where the pulse starts to rise is not practical—the amplitude, being weak and affected by noise, may not be detected. One method for cycle identification is known as the

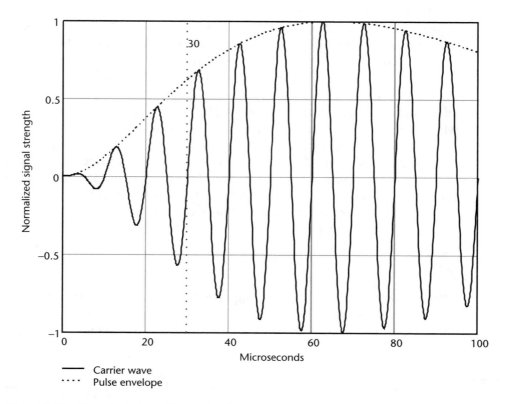

—— Carrier wave
· · · · Pulse envelope

Figure 2.26 Magnified view of Loran-C pulse.

half-cycle peak ratio (HCPR) method [8]. The time of each crossover point measured from the beginning of the pulse can be identified by the ratio of the first carrier peak before the crossover to the carrier peak right after the crossover, expressed as

$$r(t_{co}) = \frac{x_0(t_{co} - 2.5)}{x_0(t_{co} + 2.5)} \qquad (2.27)$$

The values of HCPR for several crossover points are given in Table 2.3.

When any crossover point is identified using HCPR, the arrival time at $t_{co} = 30$ μs is readily determined. A different method, using hard limiting and digital filtering, is described by Fisher [8].

While Loran-C is a hyperbolic system, based on time differences of arrival, its transmissions can be used to make time-of-arrival measurements when the target clock is very precise, within tens of nanoseconds of the transmitter clocks. This is called range-range, or rho-rho mode (see Section 2.2.3). If this mode is possible, time-of-flight measurements are made from a minimum of two transmitters, and location is found by the intersection of circles. Additional information is needed to eliminate the ambiguity occurring as a result of the circles intersecting in two places.

While Loran-C, VOR, and DME systems were first deployed many decades ago, modern navigation equipment using these methods employ the latest electronics developments in hardware and software, including advanced signal processing algorithms and displays. Therefore, accuracy and reliability have been greatly improved and equipment cost in most cases has been reduced. Even though GPS is more accurate and in many cases more convenient, there are several reasons for continuing to maintain the legacy systems. Redundancy is always desirable, and this applies to all of the described systems. Loran-C in particular has several advantages over GPS. Its signal is available in urban areas and forests for example, where satellite signals may be obscured. It also has good jamming resistance in comparison to GPS. Antenna siting is not critical with Loran-C, whereas a GPS antenna must be in view of a number of satellites spread out in the sky. In spite of these considerations, Loran-C, as well as the other classical navigation aids, will probably be eventually phased out when the authorities and agencies charged with maintaining the infrastructures behind them determine that continual reduced usage no longer justifies the costs.

2.3.4 GPS

GPS is without a doubt the most important development in location technologies from the last decade of the twentieth century. Its use encompasses highly varied applications, from maritime to aeronautic location and guidance systems, to cellular

Table 2.3 Half-Cycle Crossover Point Ratios

t_{co} μs	5	10	15	20	25	30	35	40	45	50
$r(t_{co})$	−0.13	−0.42	−0.595	−0.706	−0.781	−0.835	−0.876	−0.908	−0.934	−0.955

phone emergency assistance and the finding of wayward children in an amusement park. Its accuracy is continually being improved, and complementary and competitive systems—specifically Glonass and Galileo—provide redundancy and even better worldwide coverage. Glonass is a Russian global satellite navigation system not completely operational at this writing. The advanced Galileo satellite navigation system is sponsored by the European Community and was planned to be operational in 2008 [9]. This section gives an overview of the principles of GPS. Details of the operation of spread spectrum location systems, on which GPS is based, are provided in Chapter 3, and details of finding location coordinates from distance measurements are found in Chapter 7.

GPS is a time-of-flight–time-of-arrival distance measuring system. It determines position and velocity in three dimensions relative to global coordinates and accurate time according to universal coordinated time (UTC). It is a unilateral system where the target receiver calculates its own position by analyzing signals from spatially distributed transmitters. The GPS receiver must measure distance to at least three transmitters in order to calculate latitude, longitude, and height. When distance to one transmitter is known, the receiver position lies on a sphere whose radius is that distance. The intersection of three spheres, each having as radius the distance to one of the three transmitters, is a point whose coordinates are those of the receiving antenna. When the receiver clock is not synchronized to the transmitter clocks, as is almost always the case, a distance measurement from an additional satellite is needed to obtain a receiver clock correction value, or clock bias.

The U.S. Navstar GPS system is described here [10–14]. It comprises three segments: space, control, and user. Transmitters are located in satellites in orbit around the Earth. The full constellation began operation in 1993. An active constellation includes 24 transmitters, with additional orbiting satellites ready as spares. The nearly circular orbits are at an altitude of 20,200 km and each satellite completes its orbit in 11 hours and 58 minutes. The plane of a satellite orbit intersects the plane of the equator at an angle of 55°. There are six orbital planes, spaced equally around the equator and crossing it at longitudes 60° apart. Each plane contains four operational satellites, and on the average eight satellites are in radio view at any point on the globe at any time. Figure 2.27 is an illustration of a GPS constellation.

Frequencies of operation are 1,575.42 MHz, referred to as L1, and 1,227.6 MHz, referred to as L2. Modulation is binary phase shift keying, transmitted as a spread spectrum signal on two channels, with chip rates of 1.023 Mbps on L1 and 10.23 Mbps on L1 and L2. The chip rate is the rate of coded bits, or chips, that are modulated with the data to spread signal bandwidth beyond that required for the data alone. Navigation data that the receiver needs in order to calculate position is transmitted at a rate of 50 bps. Code and data modulation of the satellite signal on the two carrier frequencies are illustrated in Figure 2.28. The satellites use sets of orthogonal codes at the two chip rates for code division multiple access (CDMA) so that transmissions on the same frequency by all satellites do not interfere. The 1.023-Mbps code, which has a period of 1 ms, is called the coarse acquisition code (C/A-code). It is available to all GPS users and provides a less accurate positioning and timing service than is provided by the 10.23-Mbps code. The 10.23-Mbps code is called the precision code (P-code) and its period is one

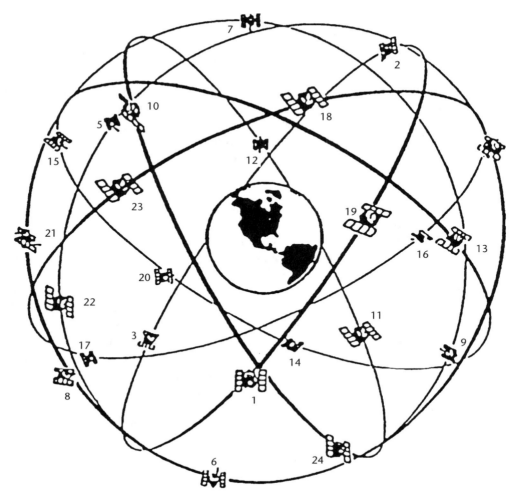

Figure 2.27 GPS satellite constellation. (*Source:* [11, p. 1-3].)

week. The P-code may be encrypted, in which case it is called the Y-code. Authorized users, notably U.S. and NATO military organizations, have access to the encryption keys and therefore can take advantage of the higher position resolution that the higher code rate, and thus bandwidth, can provide. The 50-bps navigation data is modulated on both codes. On the L1 frequency, the two modulated signals are transmitted in quadrature.

The coordinate system used for the basic position calculation is Earth centered, Earth fixed (ECEF), whose origin is the mass center of the Earth. The x-axis points to the intersection of the equator and the 0° longitude line passing through Greenwich, United Kingdom. The z-axis points to the North Pole. The y-axis is directed through the center of the Earth to the 90° longitude point on the equator. The satellite positions in space in the calculation of the receiver position are given in terms of these coordinates. We can get a feeling for the required degree of measurement accuracy needed and the size of the numbers involved as follows. The average radius of the Earth is 6,371 km, and adding the distance of a satellite

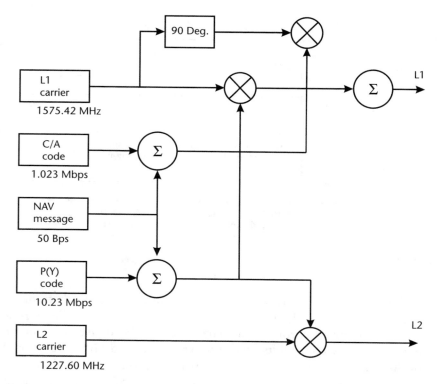

Figure 2.28 GPS satellite signal modulation.

from the Earth's surface we get 26,571 km—the approximate radius of the satellite's orbit. The lateral position accuracy of a position measurement is around 10m, and we assume we want a resolution of 1m. A binary expression to give a range along these limits would have to have at least $\log_2(26.571 \times 10^6) = 25$ bits. Time resolution for 1m = 3.3 ns. Transit time from a satellite at zenith to Earth receiver is approximately 20,200 km/3 × 10^8 m/s = 67 ms.

In order to measure time of flight, the clocks in the transmitters and receiver should be synchronized. The satellite clocks are highly accurate cesium and rubidium instruments but they are not physically synchronized and instead timing corrections are given in a data message that is contained in the signal of each satellite. This data is updated daily with corrections from ground monitoring and control stations. The GPS receiver clock, however, has much lower basic accuracy and its free-running time must be adjusted to give accurate distance measurements. Because the transmitter and receiver clocks are not synchronized, the distance measurements are offset from their true values and are called pseudoranges. The clock error, or bias, is common to all range measurements and can be canceled out by adding a fourth satellite to the measuring set in order to arrive at four equations with four unknowns, x, y, z, and t. Examples of the calculations are given in Chapter 7.

The GPS receiver must know to a high precision and accuracy the position and time of the satellites that are in view and from which distance measurements are taken. This information is provided by each satellite in the data that it transmits. A data frame is shown in Figure 2.29. The data frame is made up of 5 subframes,

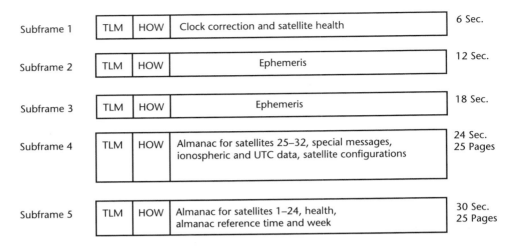

Figure 2.29 GPS data frame.

and 25 data frames make up one complete message. Each frame begins with a telemetry word (TLM) and a handover word (HOW). The telemetry word contains a preamble for synchronization. The handover word gives the exact time of transmission, in 6-second units from the beginning of a GPS week, for the beginning of the next subframe. It is used to help acquisition of the weeklong P(Y)-code. The first subframe contains data that allows the receiver to know exactly the time instant from which to measure the signal time of flight. The clocks in the different satellites are not hardware synchronized, so this data is essential in order to have a common time basis from which to measure the times of flights. The second and third subframes give the information the receiver needs to find the three coordinates of the transmitters at the point of time on the signal to which the time of flight is referenced. This information is called the ephemeris of the satellite and is given in terms relating to the properties of the elliptical orbit, called Keplerian parameters. These three subframes are transmitted during 30 seconds, and are repeated throughout the 25 frames of the message. The fourth subframe provides additional data required for measurement accuracy, specifically a correction factor for adjusting the propagation time of the signal, which changes slightly according to the parameters of the ionosphere. It also allows the receiver to convert from the satellite time, called GPS time, to universal coordinated time (UTC), so that the GPS unit, in addition to providing location coordinates, can be a very accurate real-time clock. The fourth and the fifth subframes contain almanac data from which the receiver can determine what satellites are in view in its location (provided the receiver knows its approximate location) in order to save time when searching for satellites from which it can make distance measurements. The data of the first three subframes are repeated every 30 seconds, whereas those of the last two subframes in their entirety are transmitted over 25 pages, or frame cycles. The complete data message has a duration of 12.5 minutes.

As mentioned above, the ECEF coordinate system is used for basic position coordinates, which are not suitable for most applications. These coordinates are normally translated to longitude, latitude, and height. The basis for making the transformation is the geodetic description of the Earth called WGS-84 [15]. Velocity can be computed from the change in three-dimension position over time, and/or from Doppler shift measurements on the carrier frequencies of the satellites used in the position calculations.

A key aspect of any time-of-flight distance measurement system is the ability of determining accurately a defined instant on the radio signal. Two services are available from the Navstar system: standard positioning service (SPS) and precise positioning service (PPS). Many civilian applications can make do with SPS, which gives lower accuracy but can use less expensive and possibly more compact devices with lower power consumption. PPS is used by the military and more demanding civilian applications. SPS uses the C/A short code of 1-ms length and approximately 2-MHz bandwidth, obtained from the L1 carrier. PPS uses the 1-week period P(Y)-code with a 20-MHz bandwidth. This code is sent on both L1 and L2, as shown in Figure 2.28. In addition to the fact that the larger bandwidth PPS code allows higher resolution of the time of flight, the use of both carrier frequencies allows the receiver to more accurately determine the change in signal propagation speed due to the ionosphere.

GPS uses spread spectrum phase shift modulation that overlays a binary code sequence on the carrier signal at a rate much higher than that of the data. The time of arrival that is measured in the receiver is found by finding the instant when a receiver generated version of the overlay code lines up with the incoming code signal. Details on how this is done are given in Chapter 3.

A simplified block diagram of a typical GPS receiver is shown in Figure 2.30. Code and carrier synchronization for multiple satellites in view may be carried out simultaneously by several digital tracking channels operating in parallel. The GPS receiver must make time-of-arrival measurements from at least four satellites in order to calculate three-dimensional position and time. All satellites transmit on the same L1, L2 frequencies. However, the overlay code sequence of each satellite is different and the cross correlation properties of the code effectively make satellites other than the one being tracked invisible.

Originally the location accuracy of the standard positioning service was 100-m horizontal error and 156-m vertical error 95% of the time. However, in 2000 a selected availability feature of the SPS, which allowed the U.S. Department of Defense to purposely degrade the capability provided by the C/A signal, was canceled and the accuracy improved by around two thirds. PPS horizontal accuracy is defined as 22m and vertical accuracy as 27.7m 95% of the time. Factors contribut-

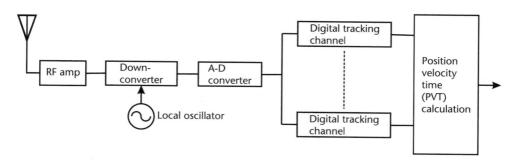

Figure 2.30 Simplified GPS receiver block diagram.

ing to errors are uncertainties in satellite time and position, and the effects of changes in the ionosphere and troposphere on propagation speed, in addition to imprecision due to receiver design regarding code sequence tracking. Another accuracy limiting factor is called dilution of precision due to the positions of the selected satellites relative to the user. The satellites should be well spread out in the sky for best results. The best position for the satellites used in the computations is for one to be directly overhead and the others equally spaced around the horizon.

GPS coordinate accuracy can be improved significantly by getting assistance from a ground station whose location is known accurately by surveying. This is called differential GPS (DGPS). The accurately located ground station makes position measurements from the satellite network and computes the differences from the known location. Correction factors are then transmitted to GPS users in the vicinity, which can use these corrections in their own calculated data to improve their location estimation. Positioning accuracy can be improved to better than 10m, and time accuracy can similarly be augmented.

While DGPS is effective when the target and the DGPS assisting terminal are no more than 250 km apart, the wide area augmentation system, WAAS, makes improved accuracy available over a continental size area, with no additional communication link required [16]. A network of 25 ground stations spread out over the United States monitor GPS satellite data and formulate correction messages for retransmission by geostationary satellites positioned to cover the continental United States. The retransmissions are sent on the GPS L1 frequency using the same CDMA scheme employed by all satellites in the Navstar constellation. Receivers equipped to use WAAS data then use the same hardware that is necessary for normal GPS location. WAAS correction can improve accuracy to within 3m. The improved accuracy is mainly due to better ionospheric corrections than are possible using the satellite navigation message when only the L1 frequency is employed. Similar wide area systems are being set up in Europe, where it is called Euro Geostationary Navigation Overlay Service (EGNOS) and in Japan, where the Multi-Functional Satellite Augmentation System (MSAS) is being developed.

For many applications, GPS receivers are overly complicated and expensive. Memory size and computing capability as well as product size and cost can be significantly reduced using assisted GPS (A-GPS). A very basic GPS receiver, consisting of antenna and RF downconversion facilities together with a relatively simple processor can obtain from a remote server the raw parameters needed for distance measurements, including:

- Precise satellite orbit and clock information;
- Initial position and time estimate;
- Satellite selection, range, and range rate.

After performing essential calculations of pseudorange and timing data, the A-GPS receiver can either process the data to find its own position, or can send the information back to the server which will perform the position calculations and will distribute the results to the necessary parties.

In addition to Navstar GPS, two other global satellite navigation systems are Glonass, developed by Russia, and Galileo, a result of cooperation by nations of the European Union. Glonass is similar in certain operational details to GPS, but it has different signal characteristics [17]. Whereas GPS uses a single-carrier frequency on each band and CDMA for multiplexing, Glonass uses a frequency division multiple access (FDMA) scheme. Each satellite frequency is defined by 1,602 MHz + k_i · 562.5 kHz on the L1 band and 1,246 MHz + k_i · 437.5 kHz on L2. k_i ranges from −7 to 13, where the subscript i refers to an individual satellite. Satellites on opposite sides of orbit operate on the same frequency. A full constellation has 24 satellites, 8 in each of three orbits inclined 64.8° from the equator and 120° apart in longitude where they cross the equator. Orbits are circular with average height above Earth of 19,100 km. Similar to GPS, there are two services: civilian and military. The civilian service uses a chip rate of 511 kHz with pseudorandom noise period of 1 ms. The military service chip rate is 5.11 MHz with sequence period of 1 second. Navigation data is modulated at a rate of 50 bps. Glonass accuracy is 57m to 70m in the horizontal plane and 70m in the vertical plane.

Development of the European Galileo navigation system commenced later than the U.S. and Russian systems, and when it becomes operational is likely to provide higher accuracy and more services and features [18, 19]. Its constellation consists of 30 satellites arranged in three planes each at an increment of 56°. Orbital height is 23,222 km. Its open service, available to all at no user cost, operates on two frequency bands of 1,164 to 1,214 MHz and 1,563 to 1,591 MHz. As in GPS, CDMA is used. Using both bands, claimed accuracy will be 4m horizontally and 8m vertically. A 15-m horizontal and 35-m vertical accuracy will be achievable when receiving on only one frequency.

The use of two or three of these systems together can result in significantly improved position accuracy. Reliability is higher, since many more satellites are visible at any location. GPS and Galileo are particularly applicable for a dual-mode receiver since they can operate on the same frequency and thus use a common RF front end.

While satellite positioning performance is constantly improving and accuracies of tens of centimeters are attainable, satellite ranging systems all have a disadvantage for many applications in that line-of-sight to several satellites is a necessity. Therefore, they will not supplant the many other location systems that have been and are continuing to be developed.

2.4 Conclusions

Factors that determine the resolution and accuracy attainable in wireless distance measuring are derived from signal parameters, system characteristics, and the physical and electromagnetic environment. High resolution ranging requires broad signal bandwidth and high energy–to–noise density ratio, coupled with a high-speed clock. Cochannel interference, multipath, and shadowing must be combated in order to realize the potential of high bandwidth and fast sampling.

Different geometric analysis techniques are used to obtain location knowledge depending on the nature of the basic measurement method—range, angle of arrival, or a combination of both. Time-of-flight location geometry depends on whether clock synchronization between the target and fixed station exist. When it does, the TOA location method finds the intersection of circles or spheres. Clock synchronization between the fixed stations is used in the TDOA method, where the target location is an estimate of the intersection of hyperbolas or hyperboloids.

The basic methods of time-of-flight location were demonstrated by descriptions of legacy navigation systems and GPS. The characteristics and operation principles of Navstar GPS and its European and Russian counterparts were also described. While GPS may eventually replace the legacy systems developed decades ago, DME, VOR, and Loran-C have been updated with contemporary digital electronics and their retention, often as a backup to GPS, contributes to increased reliability due to the diverse physical principles involved in their operation.

References

[1]   Skolnik, M. I., *Introduction to Radar Systems,* 3rd ed., New York: McGraw-Hill, 2001.

[2]   Kayton, M., and W. R. Fried, *Avionics Navigation Systems*, New York: Wiley-Interscience, 1997.

[3]   Mech, L. D., and S. M. Barber, *A Critique of Wildlife Radio-Tracking and Its Use in National Parks,* A Report to the U.S. National Park Service, February 6, 2002.

[4]   Spiegel, M. R., *Mathematical Handbook of Formulas and Tables*, New York: McGraw-Hill, 1968.

[5]   "2001 Federal Radionavigation Systems," U.S. Department of Defense and Department of Transportation, 2001.

[6]   Terman, F. E., *Electronic and Radio Engineering*, New York: McGraw-Hill, 1955, p. 1041.

[7]   *Loran-C User Handbook*, Commandant Publication P16562.5, U.S. Coast Guard, http://www.navcen.uscg.gov/loran/handbook/h-book.htm, 1990.

[8]   Fisher, A. J., "The Loran-C Cycle Identification Problem," http://www.cs.york.ac.uk/ftpdir/reports/YCS-99-318.pdf, 1999.

[9]   Galileo European Satellite Navigation System, European Commission, Directorate General Energy and Transport, http://ec.europa.eu/dgs/energy_transport/galileo/index_en.htm.

[10] *Navstar Global Positioning System Interface Specification IS-GPS-200 Revision D*, March 2006.

[11] *Navstar GPS User Equipment Introduction, Public Release Version*, September 1996.

[12] Global Positioning System Standard Positioning Service Performance Standard, Department of Defense, October 2001.

[13] Hewlett Packard Application Note 1272, "GPS and Precision Timing Applications," May 1996.

[14] Ward, P. W., J. W. Betz, and C. J. Hegarty, "Satellite Signal Acquisition, Tracking, and Data Demodulation," in *Understanding GPS: Principles and Applications*, 2nd ed., E. Kaplan and C. Hegarty, (eds.), Norwood, MA: Artech House, 2006, pp. 153–241.

[15] El-Rabbany, A., *Introduction to GPS: The Global Positioning System*, Chapter 4, Norwood, MA: Artech House, 2002.

[16] "Wide Area Augmentation System," Wikipedia, The Free Encyclopedia.

[17] *GLONASS Interface Control Document (version 5.0)*, Coordination Scientific Information Center, Moscow 2002.

[18] Hein, G. W., et al., "Status of Galileo Frequency and Signal Design," *ION GPS 2002*, Portland, OR, September 24–27, 2002.

[19] "The Galileo Project, Galileo Design Consolidation," European Commission, 2003.

CHAPTER 3

Spread Spectrum

Spread spectrum techniques involve ways of increasing the bandwidth of a transmitted data stream beyond the minimum transmission bandwidth indicated by the data rate. These techniques are prime candidates for wireless distance measuring because they provide a platform for attaining ranging accuracy which is larger than that commonly achieved on narrow band systems with comparable data rates.

In Chapter 2, two spectrum spreading methods were described: chirp and direct sequence. Other methods are frequency hopping and time hopping, as well as combinations of the four. This chapter deals with direct sequence spread spectrum. Frequency hopping and time hopping spread spectrum as applied to distance measurement are covered in Chapter 5 and Chapter 11, respectively. A system is spread spectrum if it has the following properties [1]:

- The signal's bandwidth is significantly greater than the minimum necessary to pass the data or symbol rate.
- The spreading signal that causes the increased bandwidth is independent of the data.
- Despreading the signal at the receiver is done by synchronizing and correlating a locally created replica of the spreading signal.

The main advantages of using spread spectrum techniques, compared to narrowband, are:

- Reduced spectrum power density for a given transmitted power.
- Increased immunity to jamming and cochannel interference.
- Reduced interference to other cochannel signals.
- Allows code division multiple access for concurrent use of channel by multiple terminals using the same carrier frequency.

An additional benefit of spread spectrum, which is the interest of this book, is the possibility of determining time of arrival for distance measurement to almost any degree of resolution.

3.1 Principles of Direct Sequence Spread Spectrum

A direct sequence spread spectrum (DSSS) signal is created by modulating a transmitted signal with a defined sequence of bits having a shorter duration than the data bits or symbols. These bandwidth spreading bits are called chips, and the sequence is called a code. Chip modulation may be any type, but generally binary phase shift keying (BPSK) is employed. A BPSK modulator shifts the phase of the RF carrier 180° according to the state of the modulating data bits or code chips—"zero" or "one." The data, whose symbol period is larger than the chip period, also modulates the RF signal. Data modulation is often M-ary phase shift keying, the number of phase levels depending on the data rate to symbol rate ratio. To demodulate the symbol at the receiver, first the signal bandwidth spreading by the code chips has to be canceled out. This is done by lining up a replica of the spreading code with the chips in the received signal and multiplying the locally generated and received signals together. The despreading process reduces the bandwidth of the received signal to that normally required by the data modulation, which is a function of the symbol rate. It also lowers the level of interference from cochannel narrowband signals and jammers, as well as wide bandwidth signals with different or same but unaligned spreading codes.

3.1.1 Transmitter and Receiver Configurations for DSSS

An example of a direct-sequence spread spectrum system is shown in the block diagram of Figure 3.1. BPSK is used for both data and spreading code modulation. In the transmitter the data stream phase modulates the carrier, producing a

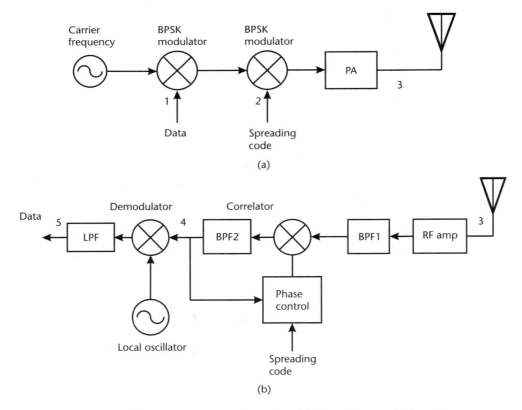

Figure 3.1 Example of DSSS transmitter and receiver. (a) Transmitter, and (b) receiver.

narrowband signal that is in turn spread by the spreading code modulation, amplified, and applied to the antenna. In the receiver the received signal and noise are amplified and filtered. Next, a correlator circuit adjusts the timing of a locally generated code sequence, identical to that of the transmitter, so that it matches the phase of the spreading code in the received signal. The meaning of phase in this context is the position of each chip in the code sequence relative to the sequence's starting point. When the local and received code sequences line up, the output of BPF2 is maximum and the resulting narrowband signal is applied to the subsequent demodulator, which reproduces the baseband data.

The spread spectrum signal as described can be expressed as [1]:

$$s(t) = A \cdot \cos(2\pi f_c + \theta_d(t) + \theta_c(t)) \qquad (3.1)$$

where $\theta_d(t)$ and $\theta_c(t)$ are the phase modulation functions of the data and the spreading code on the carrier with frequency f_c. Since binary phase shift keying is used, each phase component in (3.1) can be 0° or 180°, which is equivalent to multiplying the carrier signal by +1 or −1, in accordance with the two modulating signals. The spread spectrum signal can now be expressed alternatively as

$$s(t) = A \cdot d(t) \cdot c(t) \cdot \cos(2\pi f_c) \qquad (3.2)$$

where $d(t)$ and $c(t)$ are binary bipolar data and spreading code streams scaled to values of +1 and −1. The two forms of $s(t)$ in (3.1) and (3.2) indicate that the DSSS signal can be produced in a different manner from Figure 3.1, while giving the same result. In Figure 3.2, the logic level data and spreading code are XOR'd at baseband—the equivalent of multiplication of bipolar signals—with the logic output applied to a single BPSK modulator. Data bit transitions are normally timed to coincide with the transitions of the chips of the spreading code. When the data bit is a logic "1," the spreading code is passed to the modulator without change. When the data bit is "0," the spreading code is inverted for the duration of the bit.

3.1.2   DSSS Waveforms

The waveforms of Figure 3.3 demonstrate DSSS demodulation. Wave numbers refer to the numbered locations in the block diagrams of Figure 3.1. Wave 1 and

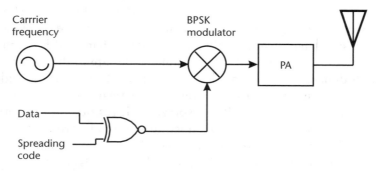

Figure 3.2   Alternate DSSS transmitter configuration.

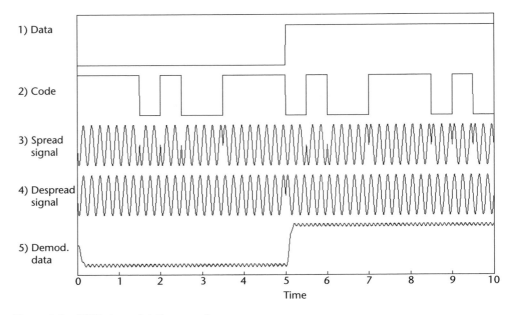

Figure 3.3 DSSS demodulation waveforms.

wave 2 are the data and the spreading code, and wave 3 is the spread spectrum signal at the transmitter and receiver antennas. Phase reversals of the carrier, equivalent to a phase shift of 180°, are evident in the transmitted signal of wave 3 at the times of transitions of the spreading code. When the data changes from 0 to 1, there is no change in carrier phase, since the change of polarity of the code signal as seen by the modulator is canceled by the inversion of the data. Multiplication of the IF signal by a bipolar locally generated replica of the spreading code that is perfectly aligned with the code imbedded in the received wave despreads the signal—collapses the bandwidth to that required by the data alone—giving the signal labeled wave 4. At the transition of dissimilar data bits the RF carrier is inverted, just as in a normal narrowband BPSK signal. This BPSK wave is coherently demodulated to reproduce the data, wave 5. The data stream at this point has a second harmonic ripple that is not completely eliminated by the lowpass filter in Figure 3.1(b).

BPF1 following the RF amplifier in Figure 3.1(b) must have a bandwidth sufficient to pass the spread spectrum signal, usually at least 10 times greater than the bandwidth required to pass a signal modulated by data alone. The ratio of the spread signal bandwidth to the data bandpass bandwidth is called the processing gain. It is also the ratio of the spreading code rate to the data rate, or symbol rate if multidimensional data modulation is used. Because of the wide bandwidth, the signal-to-noise ratio at the receiver input is relatively low, often negative (in dB). The signal-to-noise ratio after despreading is restored at the output of the second BPF to that which could be obtained in a normal narrowband system. While the despreading process does not improve the S/N in random noise, it does reduce narrowband interference by the value of the processing gain.

3.1.3 Despreading and Correlation

In order to perfectly despread the received signal, the despreading code produced in the receiver must line up exactly, that is, must be of the same phase as the code imbedded in the signal. The receiver could produce the correct phase if it knew the exact time and phase of the code in the transmitted signal, and the propagation time from transmitter to receiver. However, this knowledge is hardly ever available, so the receiver has to periodically change the phase of its replica code and test whether despreading occurs. When it does occur, the replica code phase must be maintained coincident with that of the incoming code. The ability to adjust and detect the line up of the despreading code with the received signal is the basis of the use of DSSS for high-precision distance measuring.

One way of finding out when a received signal in a direct sequence spread spectrum receiver is in phase with the replica is by using the correlation process of multiplying the local code sequence replica with the incoming signal and integrating the result. This is shown in Figure 3.1(b), where BPF2 is the integrator. A perfect transition from wave 3 to wave 4 in Figure 3.3 occurs only if the replica sequence is exactly aligned with the spreading sequence that is imbedded in wave 3. The explanation of correlation is easier if baseband signals are used, therefore we will now consider a receiver that demodulates coherently the spread spectrum signal directly to baseband, before despreading, as shown in Figure 3.4. This arrangement is consistent with (3.2) and is often applied in practice. It should be clear that the receiver architecture, with despreading in the RF chain as shown in Figure 3.1(b) or at baseband, per Figure 3.4, is in no way dependent on which spreading method is implemented, that of Figure 3.1(a) or Figure 3.2.

The spreading code is a pseudorandom sequence chosen for the property of having high correlation when matched with an exact image of itself and low correlation when matched with a time-shifted image of itself. The autocorrelation function of a periodic waveform is

$$R(\tau) = \frac{1}{T} \cdot \int_0^T x(t) \cdot x(t + \tau) \, dt \text{ for } -\infty < \tau < \infty \tag{3.3}$$

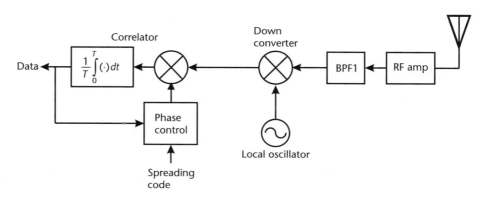

Figure 3.4 Alternate DSSS receiver configuration—baseband despreading.

where T is the period of the waveform and τ is a time shift of that waveform. Figure 3.5 is an example of a correlation function plot where $x(t)$ is a maximal length code produced by a linear feedback shift register having a period of 7 bits. Note the following characteristics of the autocorrelation function:

- It is periodic with period T.
- It is symmetric around the peaks.
- The maximum values occur at $\tau = 0$ and at integral multiples of T.

In DSSS $x(t)$ is a discrete time function consisting of a sequence of bits that are commonly called chips. These chips are not data, since their sequence is known in advance at the receiver. A "good" autocorrelation function is one that has a high ratio of peak value to the absolute value of the sidebands, which are the values of $R(\tau)$ for τ outside of the region ±1 chip around the peak. Longer sequences of a particular type of function have a higher ratio of correlation peak to side-lobes, and give a sharper indication of the point where the incoming and locally generated sequences line up.

The despreading process in a receiver is not strictly autocorrelation, but correlation of two different signals, since the received signal is not identical to the code replica generated in the receiver. The received signal contains data modulation, noise, and other interferers, and its shape is modified by bandpass filtering. Correlation is expressed as [1]

$$z_i(T) = \int_0^T r(t) \cdot s_i(t)\, dt \tag{3.4}$$

where $r(t)$ is the received signal and $[s_i(t)]$ is a set of time-displaced versions of the locally produced code sequence. Cross correlation is the correlation of two

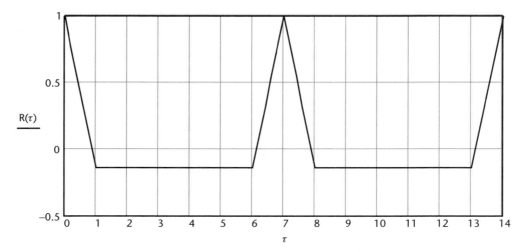

Figure 3.5 Autocorrelation of length 7 m-sequence.

signals having different codes. Cross correlation should be low for all sequence shifts so that false alignment with an unintended signal, one with a different code, is not likely to occur. Sequences $s(t)$ that have good autocorrelation properties as well as low cross correlation with other codes are chosen for despreading in DSSS receivers.

3.1.4   Code Sequence Generation

Code sequences that have good autocorrelation characteristics are often called pseudonoise (PN) sequences because their autocorrelation function is similar to that of random noise—a sharp peak and very low sidebands. For DSSS systems they must also have very low cross correlation. Two related PN families are m-sequence codes and Gold codes. In addition to their useful features, they are also easy to generate.

3.1.4.1   *M*-Sequences

A code sequence with good autocorrelation properties can be produced by a shift register with feedback taps that are logically combined and fed into the bit input. Such an arrangement is called a linear feedback shift register (LFSR). The points of the tap connections at flip-flop outputs determine the code sequence. Only certain tap connections result in the maximum length period of the sequence, which is $2^N - 1$, where N is the number of flip-flops or delay cells [2]. The sequence created is a maximum sequence, or m-sequence. For simplicity of illustration we again use the 7-chip sequence as example. First, we must define a starting point for the sequence. This is the beginning epoch of the code. The epoch is a particular reference point in the code that is agreed upon in advance. The length 7 code may be created in a three-cell shift register made up with flip-flops as shown in Figure 3.6. At a given time, logic 1s are loaded into each of the three flip-flops. Figure 3.7 shows two periods of the resulting code sequence, where the time graduations are labeled. Actually, any bit in the sequence can be defined to be the beginning. However, by loading specific bits in each cell, we can force the sequence to start at a given place at a given time. The sequence shown in Figure 3.7 starts on the first clock pulse after loading three 1s in the three cells. The loaded bits are the first to exit the shift register. This means that after the states of the flip-flops in the shift register are set by the "load" line (Figure 3.6) according to the loading

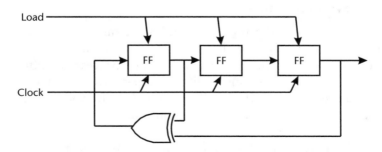

Figure 3.6   Three-cell PRN code sequence generator.

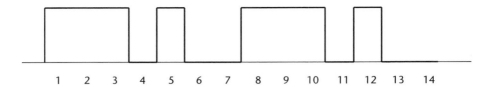

Figure 3.7 Two periods of 7-bit maximum length code sequence.

bits, it is these bits that exit the register on subsequent clock transitions, before the effect of the feedback connections appear at the output. Each of the bit transitions is a phase of the sequence of period T_s. The phase can be identified as a bit number, starting from the defined sequence beginning, as a time interval T_φ, or in angle units, where phase $\varphi = (T_\varphi/T_s) \times 360°$.

3.1.4.2 Gold Codes

While m-sequences generated from linear shift registers have good autocorrelation and cross correlation properties, only a limited number of different sequences can be produced having the same length. Gold codes are often used in DSSS systems when many codes of the same length are required, for example for code division multiple access (CDMA). Navstar GPS is an example of a CDMA spread spectrum system that uses Gold codes for its coarse acquisition (C/A) spreading sequences. Gold codes of length N are created by modular 2 adding the outputs of two m-sequence linear shift registers, each of length N. Different sequences, which have low cross correlation, are created by shifting the phase of one of the sequences, or by using different feedback. The number of N period output sequences that can be obtained equals $2^N + 1$, which includes each of the two generating sequences separately. The example of the GPS C/A code generator is shown in Figure 3.8 [3]. The outputs of two 10-bit linear shift registers G1 and G2, with different feedback combinations, are logically added to give the C/A code output. The phase of the sequence output of register G2 is determined by the connections to the register cells of the inputs S_1 and S_2. Using this arrangement, GPS creates 1,023-chip Gold codes for up to 32 satellites. The logic hardware involved in forming a multitude of codes is relatively simple, since the basic shift registers and feedback circuits are common to all of the codes.

3.1.5 Synchronization

Synchronization is the process of adjusting the clock timing and demodulation process in the receiver so that the data stream can be recognized. In a distance measuring receiver, the code synchronization process and often carrier phase locking are used to find the time of flight of the incoming signal by measuring the time shift required to line up the locally generated replica code sequence with the received embedded code. Four aspects of synchronization are:

- Alignment of a local replica code phase and rate to that of the received signal;

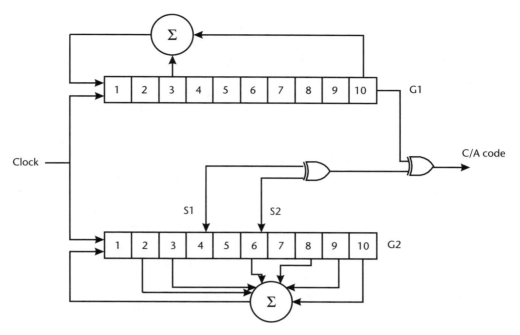

Figure 3.8   GPS C/A Gold code generator.

- Frequency and phase locking of the received carrier to the local oscillator;
- Data bit alignment;
- Data frame alignment.

There are two methods for achieving spreading code synchronization. In one, the matched filter technique is implemented by passing the signal through a series of delay lines that are polarity weighted according to the spreading code chips. Thus, the replica code resides in hardware. This method is fast, with correlation detection accomplished within two sequence periods. However, it is practical only for relatively short sequences, on the order of tens of chips. Also, it is not useful for ranging when the required time of flight accuracy is much greater than the inverse of the bandwidth of the transmitted signal. Range accuracy can be improved with matched filter despreading by averaging timing intervals over a number of repetitive measurements of the correlation peaks.

The other method of code synchronization is through use of a sliding correlator. The phase of the replica code, that is τ, in the correlation function (3.3) is varied by setting the frequency of the code generator clock to a slightly different rate than that of the received code rate, so that the replica code appears to slide slowly across the received signal code. The local and received signals are multiplied and integrated while checking for a correlation peak. Instead of offsetting the local code rate, its phase can be varied in steps whose duration is much shorter than the code sequence period. When correlation is detected, the local code rate is synchronized to the received code rate.

A frequency or phase lock loop is used to synchronize the received carrier and local oscillator. Carrier synchronization is necessary before phase demodulation can take place.

Once code despreading and carrier synchronization have been carried out, data bit transitions can be detected and the phase of the data clock can be set to the middle of each bit for reading data. The beginning of a data frame in a transmission packet is detected by correlation with a start frame deliminator or preamble sequence that is usually contained in a synchronization field at the beginning of the packet.

Synchronization is carried out in two stages, acquisition and tracking. First, during the acquisition stage, despreading to an accuracy of at least ±1 chip is carried out, to enable bit and frame alignment. Simultaneously, carrier synchronization is performed. During the tracking stage, carrier and code synchronization is constantly maintained for the duration of the transmission. Without continuous tracking, synchronization could be lost because of oscillator drift, Doppler changes due to relative velocity changes between the two stations, and signal fading.

When spread spectrum is used solely for communication, it is sufficient for the receiver to find the point of maximum correlation, where the local and received code sequences line up, so that despreading and data demodulation can be carried out. In spread spectrum distance measurement, an elapsed time must be accurately measured in order to find the time of flight of the signal. The time elapse may be the difference in time of occurrence of the same epoch in transmitter and receiver (epoch synchronization) or the time that has passed from the initialization of the code sequence (or some other given phase) and the phase at which synchronization occurs.

We saw an example of epoch synchronization in Chapter 2, Section 2.1.5, where a matched filter was employed to compress a chirp signal. The peak of a matched filter output marked an instant of time, or epoch, in the receiver that could be related to a corresponding epoch in the transmitter. In this chapter we are concerned principally with phase synchronization. In epoch synchronization, the time accuracy in marking a peak at the detector output is essentially determined by the clock rate. We discuss its use in distance measurement in Chapter 4. Phase synchronization allows obtaining much higher accuracy, which is achieved through a trade-off of the duration of the measurement.

3.1.6 Velocity Estimation

Relative speed can be measured by taking the difference between two distance readings and dividing by the time interval between them. Long tracking times limit this method to low speeds. A faster and more accurate way to measure the speed between two terminals is to find the Doppler shift, f_D, which is the difference between the transmitter frequency, f_T and the signal frequency at the receiver, f_R. The speed v is

$$v = \frac{(f_R - f_T) \cdot c}{f_T} \tag{3.5}$$

$$v = \frac{f_D}{f_T} \cdot c$$

where c is the speed of light.

The error signal in a phase lock loop is proportional to the difference between the local oscillator free-running frequency (open loop frequency) and the frequency of the received signal to which it is locked. This frequency difference includes the Doppler shift plus the transmitter frequency minus the local oscillator free-running frequency. When transmitter and receiver frequencies are not synchronized, as in most one-way systems, the line-of-sight velocity cannot be determined exactly without some additional information. Velocity can be found when Doppler shifts are measured between a nonsynchronized receiving target and multiple reference transmitters with oscillators synchronized among them. An example is GPS where the Doppler shifts are typically termed pseudorange rate measurements and include an offset due to the difference between the local oscillator frequency and the synchronized, known frequency of the satellites. The pseudorange rate, when integrated over time to give distance, is referred to as delta pseudorange. A system of at least four nonlinear equations containing as known parameters the velocity vectors of at least four satellites, the three unknown vector coordinates of the target receiver velocity, and the unknown velocity bias due to the local oscillator frequency offset can be solved to estimate the target velocity [4]. Reference [5] gives details of the measurement process for finding delta pseudorange.

3.2 Acquisition

Demodulation and data recovery of a direct sequence spread spectrum signal involves several operations:

- Carrier synchronization;
- Chip synchronization;
- Coarse despreading—signal acquisition;
- Code synchronization—code tracking;
- Data demodulation.

Signal acquisition is described next. Since there are several strategies and system configurations that can be used for acquisition, this explanation is based on the block diagram in Figure 3.9. We make the following assumptions:

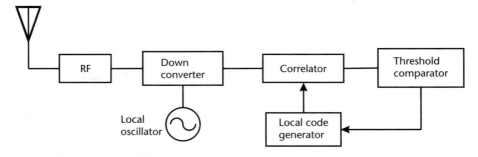

Figure 3.9 Receiver configuration for signal acquisition.

- The spreading code period is relatively short.
- Demodulation is coherent, that is, the local oscillator is phased locked to the carrier, or the IF, if downconversion is done in stages.
- Transmitter and receiver chip rates are equal.
- The code is an *m*-sequence, generated in a linear feedback shift register.
- No data is transmitted during acquisition.
- There is no noise or interference.
- Code modulation is BPSK.

The consequences of not adhering to all of these assumptions will be discussed later.

3.2.1 Code Acquisition

In order to detect data in a spread spectrum transmission, the signal first has to be despread, that is, the spreading code must be removed. To do this, the locally produced spreading code must be phase aligned with the code of the incoming signal. In Figure 3.9, the correlator receives the downconverted baseband spreading code and the locally generated code. During normal communication, it may not be important to know the time when the local code generator begins its sequence, but this time must be known for distance measurement. Figure 3.10 shows the output of the correlator as the local code generator slides the phase of the replica code in relation to the incoming signal. In this figure the phase changes by a discrete value of 1/10 of a chip per sequence period. Figure 3.10 differs from Figure 3.5 in which phase changes are continuous. Output is maximum when phases match, and decreases stepwise linearly to minimum when the phases differ by plus or minus one bit. As the phases continue to differ by larger and larger amounts, the output of the correlator remains low. The actual form of the correlator output

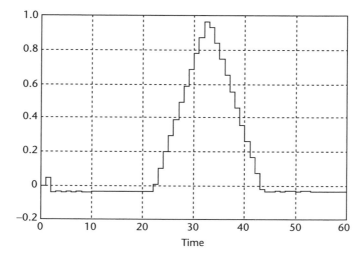

Figure 3.10 Correlator output with a code phase changes of 1/10 of a chip for each correlator integrating period.

depends on the autocorrelation properties of the particular code sequence. During acquisition, the local phase is changed by one bit or a fraction of a bit, then the correlator output is checked in the decision block for a level that exceeds a given threshold. The chip clocks of transmitter and receiver are not synchronized at this stage, and chip boundaries can differ, preventing a perfect lineup of local replica and received chips. The threshold value that is set to detect the best correlation point must take into account the worst-case skew between the transmitter and receiver clocks. The correlator presents a new output to the threshold comparator at the end of every code sequence period. If the threshold is not exceeded, the local code generator phase is changed by one chip or a fraction of a chip. When threshold is reached, the local code generator phase is left at its present value. Local code and received code are then in line to within one half of the phase change that was forced on the local code generator after each sequence period. Maximum threshold values, covering worst-case chip boundary skew between transmit and receive clocks for phase shift trials of one bit and one half bit are shown in Figure 3.11 for an m-sequence spreading code. These thresholds are

$$y_1 = (1/2)(1 - 1/N) \tag{3.6}$$

$$y_2 = (1/4)(3 - 1/N) \tag{3.7}$$

where y_1 is the threshold for 1-chip phase shifting and y_2 is the threshold for 1/2-chip phase shifting. When whole chip shifting is used, the lowest output level, y_1, occurs when the phase difference between received and replica sequences is a whole number of chips plus 1/2 chip. Similarly, half-chip shifting can achieve a maximum correlator output level of y_2 when the phase difference between the sequences is 1/4 chip.

In a real system, noise, interference, and fading affect the correlator output. Because of these effects, the correlator output could be below the theoretical worst-

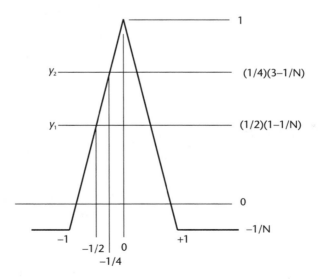

Figure 3.11 Correlator output threshold levels for 1 bit and 1/2 bit incremental shifts.

case output, and the decision circuit will not detect that code synchronization is within the coarse range and will continue phase shifting. This is called a miss. On the other hand, the correlator output may exceed the threshold at the wrong time, causing the coarse search to stop when synchronization has not been achieved. This is a false alarm. Therefore, the threshold values y_1 or y_2 should be increased or decreased according to whether a miss or a false alarm is most detrimental to system operation. A 1/2-chip increment gives a higher output on correlation and a better signal to noise ratio that will decrease the probability of a miss and of a false alarm. However, the average acquisition time is greater when a fractional increment is used, as shown below.

Figure 3.12 shows two ways of representing implementation of the correlator block in Figure 3.9. The integration form of correlation given by (3.4) is shown in Figure 3.12(a). The sample and hold element presents the results of the integration of the product of the two input sequence streams to the output where it is held for the duration of a sequence. At the end of a sequence, the integrator is reset and the signal product is integrated again. Another way of showing the accumulate and dump operation of the correlation is illustrated in Figure 3.12(b). It consists of a multiplier that receives the two bipolar sequences, a delay line of $N - 1$ cells (can be clocked flip-flops), each with a delay of chip period T_c, a summing device, lowpass filter, and sample and hold function. The circuit performs the correlation according to the discrete formulation for the continuous correlation function in (3.3):

$$r_k = K \cdot \sum_{j=0}^{N-1} x_j \cdot x_{(j+k) \bmod (N)} \tag{3.8}$$

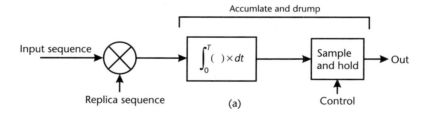

(a)

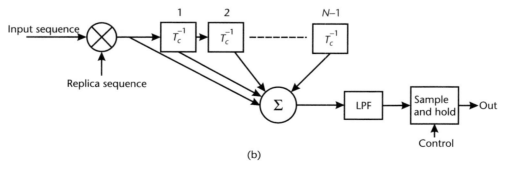

(b)

Figure 3.12 Correlator implementation. (a) Continuous configuration, and (b) discrete configuration.

where k is the phase difference in number of chips, j is the position of the chip, x_j is a bipolar chip value, and K is a scaling factor. The LPF in Figure 3.13 is typically an FIR digital filter with 6-dB cutoff frequency $1/2T_c$ (T_c is the chip period). The sample and hold control line outputs the correlation result once per sequence period, NT_c.

The threshold comparator block in Figure 3.9 makes a decision once every sequence period $T = NT_c$. If the correlator output is below the threshold, the code generator phase is increased (or decreased, depending on system implementation) by one chip, or a fraction of a chip if so designed. Otherwise, when the output is at or above the threshold, acquisition has been obtained, and the system starts the tracking mode where phase is fine-adjusted for closer correlation and synchronization is maintained during data demodulation.

In case the received code happens to be within the coarse acquisition phase difference from the replica code, tracking may commence immediately after one complete sequence period from the beginning of the acquisition mode. However, if the incoming sequence lags the replica by n chips, then n sequences will have to be tested until coarse correlation is detected (assuming one chip phase decrement each time). If 1/2-chip increments are used, the maximum duration of acquisition mode will be $2n$ sequences. Misses or false alarms will cause additional delays. Data demodulation in the DSSS system cannot commence until coarse acquisition has been obtained, so the system message protocol has to take into account the maximum coarse acquisition time. The message may have a preamble, during which the chipping sequence is sent without data. If this preamble must be kept short, to increase data throughput for example, parallel correlators may be used to reduce the time needed to check all phase difference positions. An arrangement where N correlators are used is shown in Figure 3.13. Since all possible phase shifts are tried at the same time (using 1-chip shifts), a decision on which of the N phase-shifted replicas of the code should be selected is made at the end of only one sequence by comparing the outputs of the individual correlators and selecting the maximum. The different shifted sequences are taken from taps on the local code

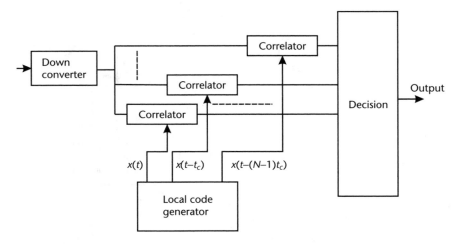

Figure 3.13 Parallel correlators.

generator—generally a shift register. All N shifts do not have to be used, and the number of correlators will divide the maximum number of sequences needed accordingly.

3.2.1.1 Code Rate Control

The circuit of Figure 3.9 is based on the sliding correlator concept where the phase of the locally generated code replica is varied by a control signal from a threshold comparator until its phase approaches that of the received signal code sequence by closer than one chip. The replica code is created in a LFSR (using the example of m-sequences), of which Figure 3.6 is an example, and the code rate is governed by the rate of the shift register clock. The clock rate can be sped up by inserting an additional pulse into the clock input pulse stream once during every sequence period, or slowed down by inhibiting one pulse during that period. Another way of controlling the clock rate is shown in Figure 3.14. The clock pulses for the local code shift register are generated in a pulse generator driven by a VCO, and divided by the number of pulses in the code sequence, N, to create a control signal. This control is used for the correlator sample and hold (Figure 3.12). It also triggers a second pulse generator in Figure 3.14, which outputs a pulse of width Δt. This pulse switches a voltage pulse $V_{\Delta t}$ into the VCO frequency control line causing a brief frequency change of the VCO output. The result is to cause the phase of the replica sequence to jump by plus or minus 1 chip during a sequence period. The relationship between frequency and phase is

$$\Delta\phi = 2\pi \cdot \int_{0}^{\Delta t} \Delta f \cdot dt \tag{3.9}$$

where Δf is the incremental VFO frequency during a period of Δt. In this case, the phase difference $\Delta\varphi$ is created by an abrupt VFO frequency jump $\pm\Delta f$ during an interval Δt, so from (3.9) the phase change in radians is:

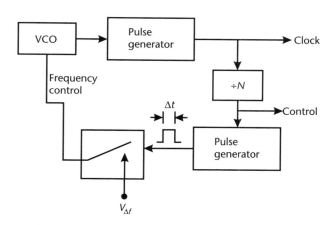

Figure 3.14 Code rate control by pulsing VCO control line.

$$\Delta\varphi = 2\pi\Delta f\Delta t \qquad (3.10)$$

The required phase is one chip, or 2π radians in terms of the chip rate, so:

$$2\pi = 2\pi\Delta f\Delta t \qquad (3.11)$$

$$\Delta f = 1/\Delta t$$

Δf can be either positive or negative for phase lead or lag. The value of $V_{\Delta f}$ in Figure 3.14 is determined by the VCO control sensitivity, k_v, in Hz/volt, so

$$V_{\Delta f} = (1/k_v)\Delta f \qquad (3.12)$$

In this example, $\Delta\varphi$ is a whole chip, but phase increments of a fraction of a chip, for better code synchronization during acquisition, can be obtained by choosing $\Delta\varphi < 2\pi$. Δt is generally chosen to be one chip period, T_c, although it can be any time span up to the sequence period, T_s. When $\Delta t = T_s$, the replica code sequence slides smoothly against the received sequence during the test of correlation.

The VCO block in Figure 3.9 can be implemented by a frequency synthesizer referenced to the system clock. In this case the designated VCO control input would digitally switch the synthesizer divider to accurately change the clock frequency by the desired amount. Another implementation based on a numerically controlled oscillator (NCO) is described below in Section 3.3.

3.2.2 Carrier Acquisition

Normally, when acquisition is attempted, the receiver oscillator and received signal are not at the same frequency and the correlation process must contend with a noncoherent signal until carrier phase lock is attained. There are two major reasons for differences in frequency between two terminals. First, no matter how accurate the oscillators are, this accuracy is not absolute, and frequency and phase differences will always be present unless they are connected together by some means of synchronization. Most communication devices have crystal time bases whose accuracy is on the order of one part per million, or one hertz deviation per megahertz of nominal frequency. Second, if the two terminals are in relative motion, the Doppler effect will raise or lower the received frequency depending on whether the distance between terminals is decreasing or increasing. The phase of the received carrier cannot be known if distance is not known, because of the wave propagation. Coarse synchronization can be accomplished in this situation by using the arrangement shown in Figure 3.15. The downconverter produces both in-phase (I) and quadrature (Q) outputs. When the local oscillator is phase locked (which implies also frequency locking) with the received carrier, the BPSK chip sequence appears only on the I output. If the frequency is locked, but the carrier and local oscillator phases differ by a fixed value θ between 0° and 360°, the spreading code modulation will be seen on both I and Q lines, with magnitudes on each proportional to

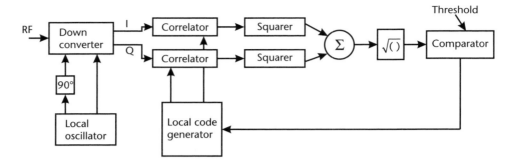

Figure 3.15 Complex signal correlation.

$$\text{Mag}(I) = A \cos \theta \tag{3.13}$$

$$\text{Mag}(Q) = A \sin \theta$$

where A represents the relative signal magnitude.

Taking the correlation of the I signal and the Q signal doesn't change the relative magnitudes of the quadrature components, so squaring the correlator outputs and summing them, as shown in Figure 3.15, gives the square of the signal magnitude:

$$(A \cdot \cos \theta)^2 + (A \cdot \sin \theta)^2 = A^2 \tag{3.14}$$

Now the comparator can decide whether or not coarse correlation has been achieved, just as in Figure 3.11, and adjust the code phase accordingly.

The remedy of squaring and adding the I and Q correlator outputs holds even if the carrier and local oscillator are not locked in frequency, as long as the frequency difference Δf, including Doppler shift, is not too great. A beat frequency Δf modulates the I and Q outputs, but the correlation results are acceptable if the phase change represented by Δf is below around 30° over the period of integration or summation, T, in the correlator. Then the period of one cycle of Δf is at the least $(360/30)T = 12T$ and $\Delta f = 1/12T$. As an example we will take $T = 1$ ms (the period of GPS C/A code). Then $\Delta f = 1/12$ ms $= 83$ Hz. If the carrier frequency is 1,500 MHz, the worst-case accuracy of the oscillators with no Doppler shift should be around $1/2 \times 83/1,500 \times 10^6$ or 36 ppm.

3.2.3 Code Rate Matching

A requirement for effective correlation is that the chip rates of transmitter and receiver be equal, or very close. How close they should be can be estimated as follows. We will assume the requirement that if the first chip of the received and replica sequences are perfectly lined up, the last bit in the integration period will be misaligned by half. Then the relative accuracy of the chip rates is $1/2N$ where N is the number of chips in the integration period. If N is on the order of thousands of chips, we see that the required accuracy is much less than that required for the carrier frequencies. Since the chip rate time base is almost always a crystal oscillator, in many cases the same as that used for the carrier frequency, it is evident that time base relative accuracy is not a significant issue.

3.2.4 Effect of Data Modulation on Acquisition

The presence of data modulation during acquisition can distort the output of the correlator although it should not prevent acquisition from being achieved. If a matched filter is used during acquisition and the data or symbol period is an integer number of code sequence periods, the data will not affect acquisition when the absolute value of the correlator output is taken for comparison with a threshold. For distance measurement when a correlator is used as shown in Figures 3.9 and 3.14, it would be preferable to suppress modulation during a message frame preamble. When the bit period is many times larger than the code period, the effect of the data is minimal. Such is the case for the C/A code of GPS where the code period is 1 ms and the data period is 20 ms.

3.2.5 Radiometric Detection

In the preceding signal acquisition description based on Figure 3.9, despreading is performed at baseband, and carrier frequency lock to the local oscillator is required before effectively correlating the received code sequence. The correlation is even more difficult in the presence of data modulation. In some DSSS systems, spreading code stripping is performed at an intermediate frequency instead of at baseband where signal power detection is used to provide a feedback signal for adjusting the replica code phase for coarse synchronization with the received code sequence. Such a system, called a radiometer, is shown in Figure 3.16 [6, 7]. The bandpass filter (BPF) has a bandwidth of $1/T$, where T is the length of a data symbol. Often, T equals the code sequence length, or a multiple of it. When synchronization has not occurred, the bandwidth of the signal at the BPF input is the same as that of the IF bandwidth, which is approximately the chip rate $1/T_c$ (3-dB bandwidth). The power output of the BPF will then be T_c/T times the signal power in the IF bandwidth. This power level out of the power detector is below the threshold of the comparator and the control circuit changes the rate of the code sequence generator, causing its phase to vary in respect to the incoming code. As the reference code phase approaches the signal code phase by less then one chip, the power increases because of partial correlation. When this power exceeds the threshold of

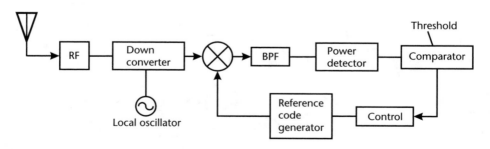

Figure 3.16 Radiometric despreading.

the comparator, the phase of the code sequence generator in relation to the incoming signal is frozen. Now the tracking procedure can be employed to decrease the phase difference between the signal code sequence and the replica sequence, thereby increasing the signal-to-noise ratio for data detection and improving the distance measurement precision as will be shown below. The performance of the arrangement of Figure 3.16 is similar to that of Figure 3.9.

3.3 Tracking

There are two aspects to tracking: carrier tracking and code tracking. Phase-shift keying demodulation requires a frequency lock or a phase lock of the received carrier, or intermediate frequency, to the receiver local oscillator frequency. A normal phase lock loop cannot be used because the carrier is suppressed in PSK modulation (however, in some systems a residual carrier is transmitted for tracking). Code tracking is necessary to suppress the spreading code, allowing the data to be demodulated through a bandpass or lowpass filter that is narrow compared to the spreading code bandwidth, and to suppress narrowband interference and other noncoherently spread DSSS signals on the same channel. First we discuss carrier tracking.

3.3.1 Carrier Tracking

Two common configurations for a BPSK demodulator are squaring loop and Costas loop. A squaring loop demodulator is shown in Figure 3.17. The IF signal is squared or doubled by other means to multiply the frequency by two. The phase is also multiplied by two, canceling the modulation since twice the 180° data shifts leaves no phase shift. The doubled frequency is locked to a VCO in a standard phase lock loop, and the VCO's output is divided by two to provide the missing carrier at the IF frequency. The reconstructed carrier can be used to provide I and Q baseband data outputs. The doubled frequency PLL does have a performance

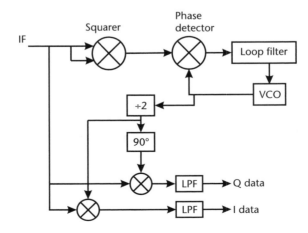

Figure 3.17   Squaring loop BPSK demodulator.

penalty compared to a PLL at the original IF, because of squaring the noise together with the carrier. The PLL in the squaring loop has to contend with at least 6 dB more noise [1]. Also, the squaring loop does not recognize the sign of the data, which is ambiguous.

Another scheme for carrier-locking a BPSK modulated signal is the Costas loop demodulator [1], shown in Figure 3.18. Performance is the same as for the squaring loop, but there is no frequency doubling. Its operation can be understood as follows. When a BPSK signal virtual carrier is phase locked to a local oscillator in a demodulator with quadrature outputs, the data appears on the I line and no signal (neglecting noise) appears on the Q output. The product of I and Q is zero, which is the voltage level applied to the VCO control. When the phase difference between the IF and VCO is not zero, the data also appears on the Q line and the product of I and Q will force the VCO into phase lock. As with the squaring loop, the Costas loop demodulator is blind to data polarity and either differential BPSK should be used, or the message protocol should provide bits to let the demodulator know the correct polarity. Both the squaring loop and Costas loop demodulators have variants for demodulating QPSK or higher levels of PSK.

3.3.2 Code Tracking

Code tracking is imperative for continuous reliable data demodulation and is the key to high-resolution distance measurement. The coarse acquisition process described above brings the incoming and local code sequences to within at least ± 1/2 chip of each other, and tracking brings them into almost perfect correspondence. One way to improve code synchronization accuracy is to vary the phase of the local code replica until the peak of the correlation tip has been recognized. This method is not particularly systematic, and the peak may not be easily recognized due to signal strength changes due to fading and perhaps motion between the terminals. A better way is to use a difference signal where a zero level indicates that synchronization has been achieved. One implementation method is the delay lock loop (DLL), shown in Figure 3.19.

The DLL uses two correlators for obtaining an error signal and an additional correlator for data demodulation. The three correlator channels are early (E), late (L), and prompt (P). This arrangement is valid only for a coherent receiver. If there is only frequency lock, and not phase lock, between the carrier and the receiver, two correlators, I and Q, are necessary for each of the channels for a total of six

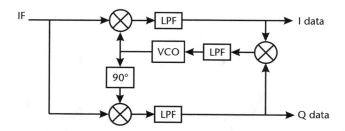

Figure 3.18 Costas loop BPSK demodulator.

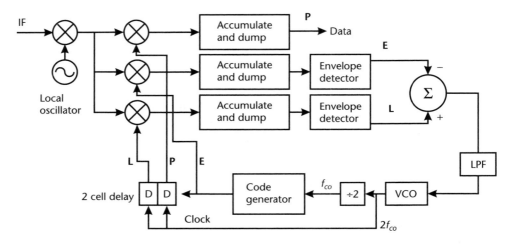

Figure 3.19 Delay lock loop.

correlators. In this case, outputs E and L come from envelope detectors, with $E = \sqrt{E_I^2 + E_Q^2}$, $L = \sqrt{L_I^2 + L_Q^2}$, and the following description still holds [5]. The terms with the I and Q subscripts are the squarer outputs of the I and Q channels in Figure 3.15.

We saw in Figure 3.5 the triangular shape of the correlation function (for an *m*-sequence), which has straight sides between offsets of ±1 bit. A curve based on the output of the envelope detectors of E and L in Figure 3.19 is drawn in Figure 3.20. The acquisition procedure described above brings the matching of the locally generated code phase to within 1/2 bit of the received signal code; that is, to approximately one-half the height of the triangle in Figure 3.20. The early and late envelope detector outputs are the result of correlation of local code generator signals with phases that are 1/2 chip earlier and later than the code generator that gives the prompt output. These signals can be expressed as [1]:

$$V_P = K \cdot \left| \frac{1}{T} \cdot \int_t^{t+T} g(t) \cdot g(t + \tau)\, dt \right|$$

$$V_E = K \cdot \left| \frac{1}{T} \cdot \int_t^{t+T} g(t) \cdot g\left(t + \tau + \frac{T_c}{2}\right) dt \right| \qquad (3.15)$$

$$V_L = K \cdot \left| \frac{1}{T} \cdot \int_t^{t+T} g(t) \cdot g\left(t + \tau - \frac{T_c}{2}\right) dt \right|$$

where $g(.)$ is the code sequence, τ is the phase difference of the prompt replica, T_c is the chip period, and K is an amplitude factor. When the prompt signal is synchronized, that is, is on time, E is 1/2 chip early and L is 1/2 chip late. The

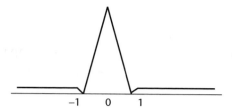

Figure 3.20 Correlation curve from envelope detector.

correlation curve is symmetrical, so the error signal is zero and the VCO frequency is not changed. However, if the prompt lags the received signal, the error line is positive and the VCO increases the replica rate to try to catch up. Similarly, if the prompt leads the incoming sequence, the error signal is negative and slows down the replica. The error signal is given by [5]:

$$error = \frac{E - L}{E + L} \qquad (3.16)$$

which cancels out the amplitude factor K. A normalized error signal curve, along with the prompt curve, is shown in Figure 3.21. We see from this curve that the linear portion in the center is less than 1 bit wide. This indicates that to prevent the code synchronization loop from losing lock at the edges, the acquisition should bring the code phases to closer than one chip of each other. In an alternate configuration, the error signal is based on power detection of E and L:

$$error_P = \frac{E^2 - L^2}{\sqrt{E^2 + L^2}} \qquad (3.17)$$

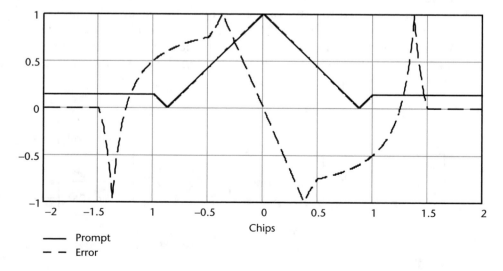

Chips

——— Prompt

− − Error

Figure 3.21 DLL Discriminator output curve and prompt output curve.

The relevant portion of this curve, between the peaks, is 1 bit wide as desired but is not perfectly linear [5].

The rate of VCO correction must be slow compared to the correlation integration period T. The rate is determined by the gain of the error feedback loop. Loop stability can be tightly controlled if a phase increment is introduced to the VCO during each integration period. A small increment can be inputted, positive or negative, after reading the error signal, and then repeated after each progressive integration period until the error is zero.

Instead of a VCO, a numerically controlled oscillator (NCO) is preferably employed. The NCO is a digital frequency synthesizer that allows exact frequency or phase increments, and its output is related to the system clock. Figure 3.22 is a block diagram of an NCO. It shows both cosine and sine digital outputs but when quadrature signals are not needed, the sine output need not be included. The inclusion of digital to analog converters followed by antialias filters makes what is often referred to as a direct digital synthesizer (DDS).

The NCO operates as follows. The phase accumulator is a binary counter with N bits that is incremented periodically by a master clock of frequency f_S. The output of the phase accumulator is plotted in Figure 3.23(a). On each clock pulse, the contents of the frequency register, $\Delta\varphi$, are added to the accumulator. The period of the generated frequency f_0 is determined by the accumulator overflows. The output frequency is

$$f_0 = \frac{\Delta\phi \cdot f_s}{2^N} \tag{3.18}$$

The phase of the output is determined by the contents of the phase register. The linearly stepped digital accumulator output can be changed to a digital sine or cosine output using look up tables in ROM. These outputs in turn are inputted to digital to analog converters, followed by antialias filters, to produce analog signals. Figure 3.23(b) shows the sine output before filtering. The output frequency is generally limited to 40% of the sampling frequency. The minimum frequency and frequency increments are $f_0/2^N$ and possible phase increments are $2\pi/2^N$. When the output of the NCO determines the chip rate, the phase increment as a fraction of a chip period determines time-of-flight increments and consequently distance resolution, which is

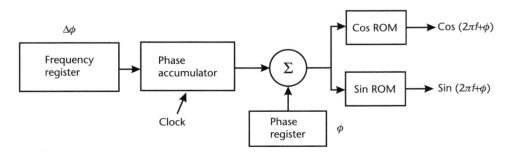

Figure 3.22 Numerically controlled oscillator (NCO).

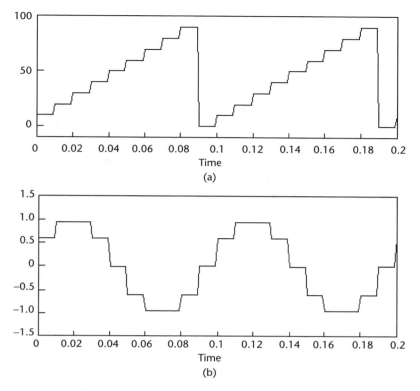

Figure 3.23 NCO output signals: (a) at output of phase accumulator, and (b) at output of sin ROM.

$$\delta d = \frac{c}{f_0 \cdot 2^N} \qquad (3.19)$$

for a one-way measurement. c is the speed of light. For example, a chip rate of 10 Mcps and an NCO with 24-bit accumulator gives a distance resolution of 2 microns! Synchronization time increases as resolution increases (smaller numbers) so attempts to get the stated best resolution are not practical. Also, noise, interference, multipath, and relative movement as well as timing inaccuracies make the actual accuracy of a DSSS distance measuring system much lower than that implied by the maximum resolution.

3.4 Measurement of Elapsed Time

Range estimation in a DSSS system is based on the measurement of elapsed time. Time is converted to distance by multiplying by the speed of propagation in the transmission path, which is approximated closely by the speed of light. The range is often not directly proportional to the measured elapsed time, which is a function of system characteristics and the method of choosing the "anchor" time from which the elapsed time is measured.

We distinguish two types of systems for measuring propagation time: one-way and two-way.

3.4.1 One-Way Systems

A one-way system measures time elapse from a known epoch in the transmitter to the time that epoch arrives at the receiver. The two terminals must use synchronized clocks. There are two methods of measuring the time elapse in a DSSS system. In one, the transmitter sends a time stamp containing the transmission time of the epoch. In the other, the receiver knows in advance a periodic data bit sequence start time that is referred to a specific real clock time, such as midnight or noon.

The time stamp measurement method works as follows. The transmitter sends a message containing the time the transmitter's code sequence begins (the time stamp). We call this time t_0. The receiver starts its replica code sequence at an arbitrary time t_1. It then measures the code shift required to synchronize with the received signal. This code shift equals τ_c. The elapsed time τ_e is then

$$\tau_e = (t_1 + \tau_c) - t_0 \tag{3.20}$$

The other way of measuring elapsed time between transmission and reception of a code epoch does not require an explicit time stamp. It is known in advance that the transmitter code sequence begins at an integral number of sequence periods from a specific clock time, t_0. Thus, the transmitted code begins at

$$t_1 = t_0 + iT_s \tag{3.21}$$

where i is a positive integer and T_s is the sequence length. The receiver begins its code clock at the same time. Then the elapsed time between transmit and receive equals the measured code phase shift at the receiver.

Navstar GPS is an example of a one-way spread spectrum system. It uses the second method described above of starting its code sequence referenced to a real time. A time stamp, based on a real time clock with a period of 1 week, is transmitted by the satellite for the purpose of helping the receiver to acquire the GPS long P(Y) code. This code has a sequence length of 1 week, so the time stamp, with a resolution of 1.5 seconds, lets the receiver adjust its replica phase to near the received code phase and thereby reduces the correlation search time. The receiver clock is not directly synchronized to the transmitter (satellite) clock so the distance measurement is a pseudorange that is used to find a precise range and location in conjunction with elapsed time measurements from a number of satellites.

In some one-way systems the transmission time is not known at all, and the transmitter may not even be a knowing party to the distance measuring procedure. In this case, the reference time is arbitrarily set by a location-estimating controller, and multiple geographically dispersed receivers are required to find distance or location. Distance cannot be found directly from elapsed time, but location is determined from time differences measured at the receivers. This is the TDOA method described in Chapter 7.

3.4.2 Two-Way Systems

A two-way system measures the out and back propagation time of a signal, and the distance between the two terminals is one-half the total propagation time times the speed of light. The initiating terminal, the interrogator, does not need to know the clock time of the responding terminal.

A two-way system works as follows, according to Figure 3.24. A spread spectrum signal is sent from an interrogator to a transponder, designated responder. If the system is operating in a duplex mode, where transmission and reception are simultaneous over separate frequency bands, the responder retransmits the signal to the interrogator simultaneously with its reception. If the mode is half duplex, retransmission is delayed to allow time for the whole packet to be received and for the interrogator to change over to receive and the responder to change over to transmit. The interrogator compares the phase of the demodulated code to that of the transmitter code generator, which runs continuously. The phase difference between transmitted and received code sequences, minus the delay, indicates the two-way time of flight from which the distance can be obtained. This two-way system will be described in detail later in this chapter. A two-way system may be used for RFID, where the responder is an inexpensive device with minimal electronic circuitry that only needs to receive, store, and retransmit the incoming signal.

3.4.3   The Time Measurement Process

In code phase synchronization the code sequence is used as a measuring stick, with individual chip boundaries serving as graduations. When a signal is received from a terminal to which distance is to be measured, the phase of the locally generated code sequence must be adjusted to correspond to the phase of the code in the received signal. One way to do this is to speed up or slow down the local code rate until a phase match is detected by the correlator. The elapsed time is measured by maintaining two running versions of the replica code sequence and comparing their phases. Both sequences start out together at the same epoch, say, the point where all 1s are loaded in the shift register (Figure 3.6). The reference code sequence that modulates the interrogator transmitter keeps running at the standard clock rate that is common to the interrogator's transmitter and receiver. The relative position of the other sequence is varied by changing the code rate or by skipping or inserting chips until the correlator indicates a match with the code of the received signal. The elapsed time is the lag in the number of chips of the variable sequence

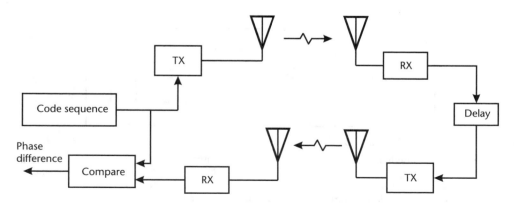

Figure 3.24   Two-way elapsed time ranging configuration.

compared to the reference sequence times the chip period. For example, in Figure 3.25 the received sequence lags the variable sequence by two chips. The time delay is measured by counting the number of chips that were inserted in the variable sequence until it lines up with the sequence of the incoming signal, and multiplying by the chip duration.

Several issues should be noted when using this method. (1) The longest time delay that can be measured equals the period of the sequence. For a longer delay, the time marks repeat themselves and there is an ambiguity about delay, and correspondingly, the distance. If additional information is available, such as an approximation of the distance or time, or a time stamp in the message that is related to an identifiable data bit transition, then the ambiguity can be resolved. (2) The time resolution is the length of a chip. (3) The clocks for the sending sequence and the reference sequence in the receiver must be running at the same rate once a correlation match has been detected. (4) Varying the sequence position by adding or subtracting chips must be done relatively slowly, preferably no more frequently than one chip out of the sequence length, so that the correlator integrator will have enough time to accumulate the result of the shift.

Another way to measure the delay is by measuring the accumulation of changes in the clock rate of the searching code sequence from the time its beginning epoch is set to the time that the correlation peak occurs when the received sequence is in synchronization with the searching sequence. We can see how this may be done through an analogy as follows. Two cars A and B are traveling the same speed and direction on the same road but B lags A by a distance d_0. B knows the way and A doesn't want to get lost, so A reduces speed until B starts to overtake A. How can A find out the original distance between them by observing the speedometer and a clock?

B's speed is constant, v_B. Once A starts to slow down at time t_0, his speed is a function of time, $v_A(t)$. The distance between them is $d(t)$, which can be expressed as

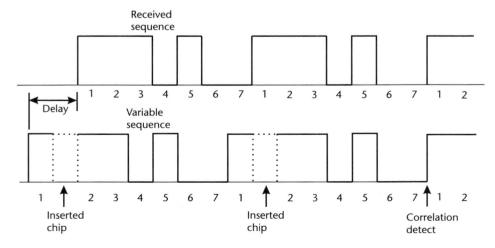

Figure 3.25 Elapsed time of received signal sequence.

$$d(t) = d_0 + \int\limits_{t_0}^{t} [v_A(\tau) - v_B] \, d\tau \tag{3.22}$$

When the two cars arrive side by side, $d(t) = 0$, and A speeds up to the same speed as B. While A continues to try to maintain his position parallel to B the integral, whose value is known, will give the distance that separated the two cars when A started trying to let B catch up:

$$d_0 = \int\limits_{t_0}^{t} [v_B(\tau) - v_A] \, d\tau \tag{3.23}$$

In terms of the distance measuring spread spectrum receiver of Figure 3.26, B is the code sequence in the incoming signal, v_B is its chip rate. A is the replica code generator output in the receiver, and v_A is the chip rate corresponding to the rate of the reference clock generator, which advances the output one chip per clock pulse. The integral of the clock rate, or 2π times the frequency of the oscillator that drives the clock in terms of radians, is the phase of the replica and reference sequences. The integral could be implemented as shown in Figure 3.26. A replica pulse counter counts (integrates) the pulses connected to the replica clock input, whose rate is continuously adjusted to achieve and then maintain correlation with the incoming signal. A reference pulse counter accumulates regularly spaced pulses at the system chip rate, which is the same as the rate of the received sequence. Both counters are reset at the same time—that of a given epoch, which could be the time of start of a code sequence at the transmitter. The difference of the readings of the two counters, modulo the sequence length, is the phase difference in bits between the two sequences at the time the counters were reset. This difference is multiplied by the bit period to get the time of flight and then translated to distance by multiplying by the speed of light.

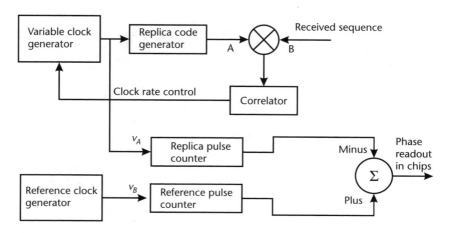

Figure 3.26 Phase difference indicator implementation.

3.4.4 High-Resolution Elapsed Time-Measuring Receiver

Figure 3.27 is a block diagram of a distance measuring spread spectrum receiver that measures elapsed time to a small fraction of a chip period. The spreading sequence and the data are modulated using PSK—usually BPSK for the spreading sequence and BPSK or QPSK for the data. The individual blocks were analyzed separately in the preceding discussion, and now we shall see how they are put together. The RF amp represents the receiver front end, which is designed for the operational frequency of the communication link. This block usually includes a downconversion stage so that most amplification is performed at an intermediate frequency. The IF bandwidth must be wide enough to pass the spreading signal, so the signal-to-noise ratio into the mixer is relatively low compared to a narrowband receiver that demodulates directly the transmitted data. The mixer and carrier NCO local oscillator source convert the received signal from IF to baseband. The output of the mixer is quadratic, and therefore the lines into the three correlator blocks are each I/Q pairs and each block consists of two correlator functions. Each I/Q correlator block correlates the baseband signal with a prompt, early, or late replica sequence and outputs the I/Q result to function blocks shown on the diagram. The prompt outputs are used for carrier tracking, which may be a Costas loop as described above, code acquisition, and data detection. The control signal from the Costas loop in the carrier-tracking block adjusts the frequency and phase of the carrier NCO for coherent PSK demodulation. Not shown in the diagram is an AGC signal that is produced from a prompt correlation envelope detector and is applied to the RF amp gain stages.

The prompt, early, and late correlator outputs are used for acquisition and tracking. During acquisition, the code NCO phase is periodically adjusted until the envelope of the prompt correlator output exceeds a threshold. At this point, the tracking function begins, with the code NCO receiving its phase control from the code phase discriminator in the acquisition/tracking block. The code NCO

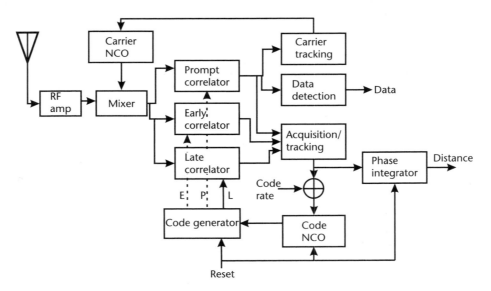

Figure 3.27   Spread spectrum distance measuring receiver.

output serves as the clock for the code generator. The clock rate is determined by an input to the NCO frequency register (Figure 3.22) at the nominal code rate to which is added the output from the acquistion/tracking block. A two-times code rate clock is also input to the code generator to produce the P and L phase-shifted code sequences (Figure 3.19).

The distance measurement is obtained as follows. The code generator phase is set to a reference epoch, such as all 1s in the LFSR, and the NCO phase accumulator is set to 0. The time of this reset signal is the known or estimated time, at the receiver, of the transmission of the signal from the terminal to which distance is to be measured. Feedback phase corrections from the acquisition/tracking block to the code NCO are accumulated in a code phase integrator and the result is read out after acquisition and during steady state tracking. The elapsed time is the accumulated phase in degrees divided by 360 and multiplied by the chip period. The transit time (TOF) must be less than the code sequence length to give an unambiguous result. The distance is the TOF times the speed of light.

The system just described can be used for accurate positioning based on time-of-arrival (TOA) and time-difference-of-arrival (TDOA) location methods. In a unilateral configuration, such as GPS, a one-way TOA is estimated based on known time of transmission and measured TOF of an epoch of a sequence. The transmitter sends a time stamp of its transmission to the receiver. The epoch that relates to this time stamp is known. For example, a GPS satellite sends its time in a framed data message, and the receiver knows that this time pertains to the next subframe crossover after the time stamp. The receiver then knows the instant of a sequence epoch and it resets its code generator to this epoch and simultaneously resets the code NCO accumulator. The accumulated phase of the NCO when the terminal is tracking the received code can be translated to one-way time of flight and to the distance between the terminals. A GPS receiver time base is not synchronized to that of the satellite so the measurement that is made is called pseudorange, which differs from the actual range by an amount attributed to the offset of the receiver clock from the satellite clock. The clocks of all satellites are known accurately from navigation messages, so the clock offset between the receiver to all satellites is the same and can be recovered by solving simultaneous equations whose unknowns are the three receiver location coordinates and the clock offset.

A multilateral configuration, consisting of a transmitting terminal whose location is to be determined and multiple receivers in the region where the transmitter is expected to be, performs position estimation based on the TDOA method. The clocks of all receivers are synchronized. The transmitter sends a spread spectrum beacon asynchronously. All receivers reset their local code generator epoch and NCO accumulator at a given instant. The location is then found from the solution of time difference of arrival simultaneous equations (Chapter 7).

3.4.5   Duplex and Half Duplex Two-Way Ranging Examples

Following are two examples of relatively simple ways to determine distance between two wireless terminals. The distance measured is derived from one-half the time of flight from an interrogator terminal to a responder terminal and back. The particular protocol and hardware depends on whether the communication link is duplex or half duplex. The figures do not include the data modulation.

3.4.5.1 Duplex Two-Way Ranging

Figure 3.28 shows a distance-measuring system block diagram for a duplex link. Transmission and reception at each terminal occur simultaneously on two different frequencies, which are separated enough so as not to interfere. The interrogator transceiver has two code generators, NCO clock sources, and frequency synthesizers. All frequencies are derived from the same reference oscillator. Both NCOs operate at the same clock rate and are synchronized at the beginning of the interrogator transmission by a reset command that sets the transceiver transmitter and receiver code generators to the first code epoch and resets the NCO phase accumulators. The responder receiver down converts the incoming signal to IF, bandpass filters it, then amplifies and upconverts the signal to a frequency that differs from the receiver frequency by Δf. Retransmission is simultaneous with reception, so the responder acts as a mirror to the interrogator transmission. At the interrogator, round-trip time-of-flight measurement takes place as described above for Figure 3.27. The phase of NCO 2 is varied to update the code phase according to the correlation output in the acquisition/tracking unit and when code alignment occurs, the accumulated phase since reset is read out in the phase integrator. The two-way time of flight t_{2W} is

$$t_{2W} = (\varphi/360)\, T_C \qquad (3.24)$$

where φ is the accumulated phase in degrees and T_C is the chip period. The distance between terminals is

$$d = (t_{2W}/2)\, c \qquad (3.25)$$

The responder in this example is very simple as it does not perform any baseband processing of the signal. Its weakness is that it increases significantly the inband noise. The signal-to-noise ratio of the received signal in the spreading code bandwidth is reduced by the sum of the receiving and transmitting noise figures of the responder.

3.4.5.2 Half Duplex Two-Way Ranging

A method of measuring two-way time of flight on a half duplex link is shown in Figure 3.29. It uses matched filters for correlation and has a basic resolution of plus or minus one chip. This resolution can be improved by using shift registers in the matched filters of length a multiple of the sequence length with correspondingly increased clocking rate. Also, by averaging multiple time-of-flight measurements the time-of-flight estimation can be improved. A common frequency synthesizer is used for transmitting and receiving at each terminal. Timing of the transmitted and received signals is illustrated in Figure 3.30. The interrogator transmits a burst that includes at least one code sequence, normally preceded by a preamble (not

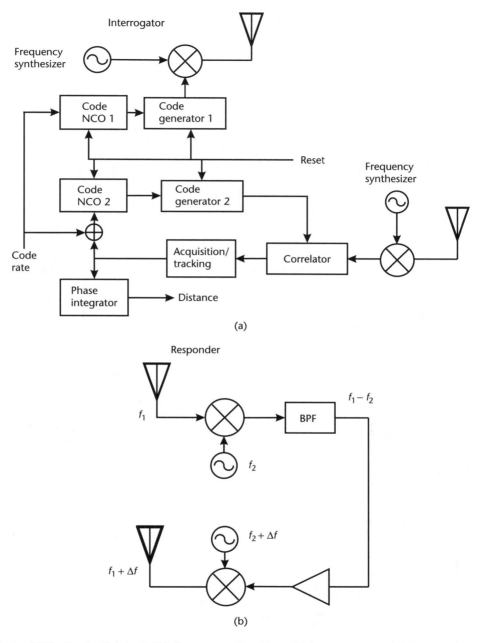

Figure 3.28   Duplex link for DSSS distance measurement. (a) Interrogator, and (b) responder.

shown) for frequency synchronization and automatic gain control (AGC). The sequence is initiated by a reset command that also resets a counter that is advanced by the code generator chip clock. At the conclusion of the sequence burst the interrogator switches to receive. When the responder detects the end of the received sequence, the matched filter output activates a command to reset the NCO and code generator, which now continue to produce code chips synchronized to the

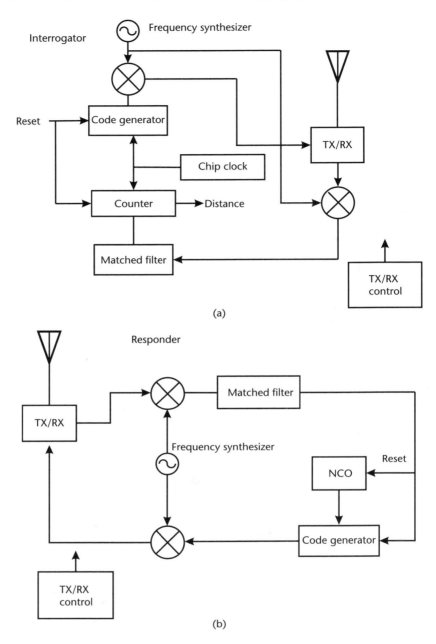

Figure 3.29 Half duplex link for DSSS distance measurement. (a) Interrogator, and (b) responder.

received code rate and phase during changeover from receive to transmit modes. The start of the responder transmission is not synchronized and will occur at any time during a code sequence. When the interrogator receives a complete code sequence, its matched filter stops the counter. The counter state is the total time, measured in chips, between the beginning of the transmitted sequence and the end of the received sequence, and includes twice the propagation time t_P. The total time T_M, in seconds, which is the counter state times the chip period, is

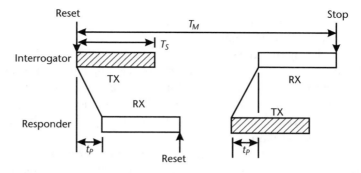

Figure 3.30   Half duplex timing diagram.

$$T_M = nT_s + 2t_P \tag{3.26}$$

where n is the total number of whole sequences. The distance between interrogator and responder is:

$$d = \frac{c}{2} \cdot (T_M \bmod T_S) \tag{3.27}$$

The local oscillators in the interrogator and responder must be stable enough so that drift between them will not be significant outside of the times that the signals are correlated by the matched filters. While the simplified block diagram in Figure 3.29 shows the counter driven by the chip clock, precision is improved by clocking it at a rate several times as high, so that the resolution of the counter will not affect the elapsed time measurement. The described method has a resolution on the order of one chip, so the chip rate must be chosen according to the required range accuracy.

3.4.6   Sequence Length and Chip Period

The sequence length and chip period have particular importance from the point of view of ranging. The sequence length T_S determines the maximum unambiguous total time of flight that can be measured. For one-way links, maximum unambiguous range is $T_S \cdot c$ and for two-way links it is $T_S/2 \cdot c$. If the range, R, is greater than $T_S \cdot c$ for one-way links, the measured phase displacement may be interpreted as a time of flight $= (R/c) \bmod T$. The apparent total time of flight of two-way links is $2(R/c) \bmod T$. For example, assume $T = 31\ \mu s$. Maximum unambiguous range for a two-way system is $(31\ \mu s \times c) \times (1/2) = 4.65$ km. An actual range of 8 km will be measured as 3.350 km.

Actually, it is possible to find the real range even if the sequence length is shorter than the time of flight if additional information is available to eliminate the ambiguity. A prime example of this is GPS Navstar. The period of the C/A (course/acquisition) code is only 1 ms, whereas the time of flight from a satellite is around 70 ms. However, the satellite message contains a time tag that gives the transmission time at a particular bit transmission in the message. The receiver can distinguish this epoch to better than 1 ms so there will be no ambiguity in the received time measurement.

Considering that the sequence period T_S is fixed, we now look at the significance of the chip duration T_c. First of all, the processing gain, which determines immunity against jamming and narrowband interference, is equal to the ratio of the symbol period and the chip period. The symbol period equals the sequence period in many systems, but not all. In any case, if the symbol period is fixed, a shorter T_c means larger processing gain.

A second point is that time-of-flight precision, and thus range, is directly proportional to the chip duration. A chip rate of 1 Mbps, for example, gives a coarse range precision of $(1/1 \text{ Mbps})(c) = 300m$. We have seen that range estimation during acquisition can be a fraction of this—generally 1/2 or 1/4—and as we have shown above, the tracking process allows a much higher precision to be attained, but still related to the chip duration.

The shorter the chip duration, the better the multipath rejection. Multipath reflections that are greater than T_c, up to $T_S - T_c$, will not be correlated and therefore will not interfere, assuming that correlation is successfully achieved on the direct signal. Some spread spectrum receivers have multiple correlators to purposely correlate reflections so that they may be combined with the direct signal to increase the total signal power and thus the signal-to-noise ratio. They are called *rake receivers*, alluding to the correlators as prongs on a garden rake.

We see that a short chip period is best for ranging and interference rejection, but the consequence is that faster clocks and high-speed digital processing is required to implement it. Signal bandwidth is also increased, but often telecommunications regulations limit the power density while allowing higher bandwidth. This is the case for UWB, which we will look at in Chapters 10 and 11.

3.5 Propagation Time Resolution

DSSS is used for high-resolution ranging due to its ability to measure to a fraction of a chip the amount of correction required to line up a received sequence with a locally produced replica. A prime advantage of the sliding correlator procedure described in this chapter is that the potential precision obtainable is much higher than the resolution of the system clock. This precision is obtained, however, by a trade-off with measurement time. TOF measuring precision is limited by several factors. Noise and multipath propagation limit the ultimate ranging accuracy that can be achieved. Performance degradation due to noise is reduced by reducing the bandwidth in the distance measuring system, again at the expense of measuring time. Combating multipath interference generally requires increasing predetection bandwidth and chip rates as echo path time differences decrease, as in indoor environments.

3.5.1 Tracking Accuracy and Noise

A fundamental limitation to the accuracy that can be attained in measuring distance is due to random noise. While the tracking discriminator may have a capability of high resolution, the distance precision limitation is determined by the signal power to noise density ratio, C/N_0, the bandwidth of the tracking loop B_L, and the early to late correlator spacing in fraction of a chip, d. It also depends on the type of tracking phase discrimator. The variance of the tracking error, σ_τ^2, in units of chips squared, for a coherent tracking loop, is [8]:

$$\sigma_\tau^2 \approx \frac{B_L \cdot d}{2 \cdot \left(\dfrac{C}{N_0}\right)} \tag{3.28}$$

Discriminator types may be classified according to the way the early and late correlator outputs are combined to create an error signal to control the rate of the oscillator that drives the code generator [5, 8]. Figure 3.19 shows a coherent delay lock loop (DLL) in which the late and early signals, L and E, are subtracted to form the error signal. In a noncoherent system in which the incoming carrier is not phase-locked to the local oscillator, there are two types of combinations of the inphase (I) and quadrature (Q) outputs of early and late correlator pairs, written as I_E, Q_E, I_L, Q_L. In one, based on early and late power signals, the error signal is

$$D_P = \left(I_E^2 + Q_E^2\right) - \left(I_L^2 + Q_L^2\right) \tag{3.29}$$

The other is a dot-product where the error signal is formed from

$$D_{DP} = (I_E - I_L)I_P + (Q_E - Q_L)Q_P \tag{3.30}$$

For the noncoherent discriminators, the tracking error depends inversely on the predetection integration interval, that is, the integration time of the correlator. All three types of discriminators converge to (3.28) for large values of C/N_0, starting from around 30 dB-Hz. From (3.28) it is apparent that the tracking error can be improved by reducing the loop bandwidth, as well as by decreasing the early to late correlator spacing d. Equation (3.28) is based on the assumption of infinite bandwidth of the signal that reaches the correlator (i.e., signal pulses are truly square). The effect of restricted signal bandwidth on the accuracy of (3.21) is more pronounced as d is reduced from unity [8]. The equation is reliable for d equals one chip and the RF signal bandwidth at the first spectrum null points is $2R_c$, where R_c is the chip rate. As d is reduced, the signal bandwidth must be increased in approximately the same proportion in order to achieve the tracking error predicted by (3.28).

An example shows the distance accuracy that may be obtained from a particular DSSS communication system.

Example 3.1   The given relevant system parameters are:

Data rate = 1 Mbps
Chip rate = 11 Mcps

Front end bandwidth = 2 Mbps
Receiver sensitivity = −100 dBm
Noise figure = 5 dB
Noise density = −174 dBm-Hz
DLL bandwidth = 1 kHz
$d = 1$

The predetection carrier power equals the sensitivity minus the noise figure: −105 dBm. $C/N_0(\text{dB}) = -105 - (-174)$ dB-Hz = 69 dB-Hz. Substituting in (3.28) and taking the square root of the result gives the standard deviation of the tracking error:

$$\sigma_\tau = 0.09 \text{ chip}$$

The chip length = 1/11 Mcps = 90.9 ns. The standard deviation of the tracking error in time units = 90.9 ns × 0.09 chip = 8.2 ns. Multiplying by the speed of light we get the standard deviation of the distance error σ_d:

$$\sigma_d = 8.2 \text{ ns/chip} \times 3 \times 10^8 \text{ m/s} = 2.5\text{m}$$

As mentioned, this result could be improved without changing system parameters by reducing the loop bandwidth. As the loop bandwidth is reduced to the order of tens of hertz, other noise factors come into play, notably Doppler frequency changes and oscillator phase noise. Doppler effects can be reduced by using rate aiding from a carrier-tracking loop [5]. The carrier-tracking loop keeps the local oscillator phase-locked to the signal. Adding the error signal of this loop, after proper scaling, to the DLL loop helps the latter to maintain the replica code rate synchronized to the code rate of the signal.

It is easy to see that designing for a high-chip rate–short chip duration for a given sequence length or data rate (i.e., high processing gain) pays off for distance measuring accuracy. For the same carrier power and loop bandwidth, the tracking timing accuracy in terms of percentage of chip length is constant, so for the shorter chip, the distance accuracy is improved.

3.5.2 Multipath

The most serious impediment to good range accuracy in short-range distance measurement, particularly indoors, is multipath interference. One important reason for using spread spectrum techniques for communication is its inherent resistance to multipath interference, compared to sending the same data rate over a narrowband channel. We have seen that the DSSS code is a pseudonoise sequence that has a sharp correlation peak and weak sidelobes. This means that an interfering signal with the same spreading code as the desired signal that is out of phase by ±1 bit or more will not affect the desired signal that has been synchronized with the local replica code. In terms of time delay, a spreading code with a high chip rate will be impervious to interference from received signals that are delayed by more than its chip period. Referring to multipath interference, reflections that reach the receiver on a path that is greater than the line of sight signal by more than the distance traveled by the wave during one chip period will not significantly affect reception.

While multipath returns that are spaced by more than a chip time from the line of sight signal may be avoided in the correlation process, it is possible that the acquisition stage in the synchronization process will try to lock on to one of the reflected signals. The range result, in this case, will be too high because the reflected path is longer than that of the line of sight path. Many spread spectrum receivers have rake correlators (see Section 3.4.6) that attempt to lock on to the individual multipath signals, and then to combine them in order to extract maximum power from the transmission and thereby improve the signal-to-noise ratio. Each correlator in a rake receiver uses a different delay of the replica code to test for arrival of the desired signal. The values of the delays in the parallel correlator channels may be established by trial and error or by making a channel impulse response estimation during a message preamble. When the preamble is known in advance, the receiver can deduce the multipath environment from an analysis of the complex spectrum that results after doing a Fourier tranform of the preamble samples. By checking the arrival times of signals that are received over different paths, the earliest signal can be detected and tracked to carry out the ranging function.

In short-range indoor communication, reflected wave delays are likely to be on the order of tens of nanoseconds and may well be shorter than the spreading code chip period. Some methods of reducing intersymbol interference to communication due to multipath may not be effective, or may be destructive, for range measurement because they may blur the time of flight. The effect of multipath on the tracking error is examined below for one multipath signal.

Figure 3.31 shows the DLL discriminator outputs from a direct line of sight signal and a delayed and attenuated multipath reflection drawn on the same scale [8]. The feedback error signal acts to force the discriminator output to zero, which occurs, considering only the direct signal, when the received signal and the replica are perfectly lined up, or in phase. The composite signal formed from the addition of the direct and reflected multipath signals, shown as a dashed line in Figure 3.31, crosses zero at a point that is bounded by the peaks of the direct signal output

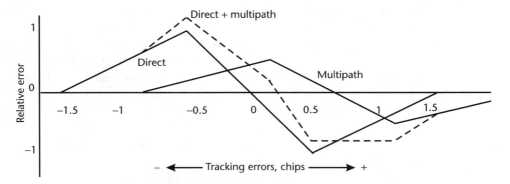

Figure 3.31 Direct and multipath outputs of DLL discriminator. The dashed line is their sum. (*From:* [5]. © 1999 IEEE. Reprinted with permission.)

and those of the multipath signal output. Thus, multipath interference causes the DLL feedback loop to achieve equilibrium when the receiver replica code is not lined up perfectly with the code in the received line of sight signal. The extent of the error in the receiver replica code phase is a function of the ratio of the multipath signal to the line of sight signal and the phase relationship between them. The error is reduced with narrow correlator receiver architectures; that is, with less than one chip between the DLL discriminator early and late codes [8].

3.5.3 Increased Range Resolution Using Carrier Phase

The accuracy of range and location coordinates of a target are increased considerably by measuring the phase offset between the RF carrier of the received signal and the receiver local oscillator, in addition to the code offset measurement discussed above. A phase comparator circuit can measure carrier phase offset to within a small fraction of a cycle, and considering that the signal travels one wavelength during a cycle, the phase measurement gives a resolution of a fraction of a wavelength. At the GPS L1 frequency of 1.575 GHz, for example, the wavelength is 19 cm. If we assume that carrier phase can be measured at an accuracy of within 15°, then the range accuracy will be (15/360)19 = 0.8 cm. This is two orders of magnitude better than the accuracy obtained by measuring spreading code displacement, which is around 2m for a GPS receiver using the precision (P) code.

The problem with realizing the potential accuracy of carrier phase ranging is that the distance between the communicating terminals is much greater than a wavelength and the number of whole wavelengths in that distance is difficult to ascertain. If N_d equals that number of wavelengths and φ equals the phase comparison result, $0 \leq \varphi < 2\pi$, then, assuming that the receiver local oscillator cycle begins at the epoch of target symbol transmission, the distance to the target is:

$$d = N_d \lambda + \frac{\varphi}{2\pi} \lambda = \left(N_d + \frac{\varphi}{2\pi} \right) \lambda \qquad (3.31)$$

The integer variable N_d is ambiguous as far as the measurement of φ is concerned, and so the value of d can be determined only to the degree that the range of possible values of N_d is known.

We have seen previously in this chapter that incoming and local replica code alignment can be determined to within a fraction of a chip. If this accuracy is within a wavelength, then the carrier phase measurement can add considerably to the range resolution. In the hypothetical situation pictured in Figure 3.32, transmitted and received code displacement is measured to within one-fifth of a chip. There are five carrier cycles per chip period, and they are synchronized to chip boundaries. After acquisition and tracking, the time of flight τ is found to be between 2.4 to 2.6 chip periods and measured carrier phase is $\varphi = 2.5$ radians. $N_d = 2.4 \times 5 = 12$. The estimated distance is, according to (3.31), $d = 12\lambda + (2.5/2\pi)\lambda = 12.4\lambda$.

Mitchell [9] has suggested a similar method for tracking the distance between satellites, where the chip rate is 300 chips/second and carrier frequency is 12 GHz. In his example, pinpointing the carrier cycle within a chip is done by making a

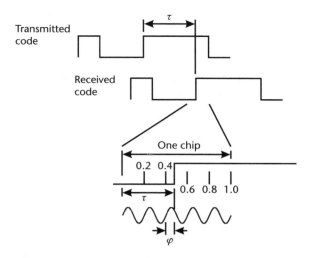

Figure 3.32 Increased time resolution using carrier phase comparison.

phase difference measurement on a 300-MHz signal using oscillators synchronized to the replica and received codes, instead of using the DLL tracking mechanism. He shows that the accuracy of this procedure is sufficient to find the correct carrier cycle among the 40 that make up the chip period.

As mentioned, a typical accuracy of a GPS receiver synchronizing the P code is 2m, whereas the wavelength at 1.575 GHz is 19 cm. This leaves a carrier cycle ambiguity of 10 cycles that must be resolved in order to take advantage of the accuracy provided by carrier phase difference measurement. Differential carrier phase positioning methods cancel out various measurement errors and eliminate the carrier cycle ambiguity. A reference receiver, whose position is known precisely and is located within tens of kilometers from the user receiver, makes carrier phase measurements while continuously tracking multiple satellites. These measurements, together with measurements made by the user GPS receiver, are processed to cancel out ambiguities. When satellite tracking is performed over a period of at least 30 minutes, the distance between the reference receiver and the user receiver can be estimated to closer than 1 cm.

3.6 Conclusions

In addition to its interference rejection properties, DSSS is particularly appropriate for distance measurement because it provides a systematic manner, through a closed loop control mechanism, of achieving high resolution from signal bandwidths that are relatively low. For example, the short GPS C/A code can provide an accuracy of 10m with a chip length of 976 ns, equivalent to a distance of 293m. Accuracy is obtained at the expense of processing time, however. While a high-speed system clock is not required for high-accuracy time difference measurements, one should be aware of the fact that TOF or TDOA measurements do require a precise measurement of the beginning of the phase difference process in the DSSS receiver in order to utilize the high-resolution code phase difference estimation.

Noise and multipath are ultimate factors in the accuracy of the range estimation. Noise can be countered by low DLL loop bandwidth, with the penalty of increased measurement time and susceptibility to disturbance by system dynamics and phase noise. Multipath interference is alleviated by using high spreading code rates and statistical estimation when location is determined by spatially separated base stations.

References

[1]   Sklar, B., *Digital Communications Fundamentals and Applications,* 2nd ed., Upper Saddle River, NJ: Prentice-Hall, 2001.

[2]   Dixon, R. C., *Spread Spectrum Systems,* 2nd ed., New York: John Wiley & Sons, 1984.

[3]   *GPS Navstar, Global Positioning System Standard Positioning Service Signal Specification,* 2nd ed., US Coast Guard Navigation Center, June 2, 1995.

[4]   *NAVSTAR GPS User Equipment Introduction, Public Release Version,* DOD Joint Program Office, September 1996.

[5]   Ward, P. W., J. W. Betz, and C. J. Hegarty, "Satellite Signal Acquisition, Tracking, and Data Demodulation," in *Understanding GPS: Principles and Applications,* 2nd ed., E. Kaplan and C. Hegarty, (eds.), Norwood, MA: Artech House, 2006, pp. 153–241.

[6]   Nicholson, D. L., *Spread Spectrum Signal Design,* Rockville, MD: Computer Science Press, 1988.

[7]   Peterson, R. L., R. E. Ziemer, and D. E. Borth, *Introduction to Spread Spectrum Communications,* Upper Saddle River, NJ: Prentice-Hall, 1995.

[8]   Braasch, M. S., and A. J. Van Dierandonck, "GPS Receiver Architectures and Measurements," *Proceedings of the IEEE,* Vol. 37, No. 1, January 1999.

[9]   Mitchell, G., "High-Accuracy Ranging Using Spread-Spectrum Technology," *15th Annual AIAA/USU Conference on Small Satellites,* Logan, UT, August 13–16, 2001.

CHAPTER 4

Time Transfer

The most straightforward method of wireless distance measuring uses synchronized clocks in the initiator and the target. Only a one-way ranging ranging link is required. The target transmits a recognizable event with a time stamp and the initiator receiver notes the time of arrival. The time of flight is the receiver time of arrival of the event minus the time stamp. Virtually none of the applications that are considered in this book use synchronized clocks on all terminals, although multilateral networks used for location do have them on the fixed terminals. One-way time transfer to synchronize clocks is not possible when the distance between the terminals is unknown. However, using two-way time transfer methods both clock time difference and propagation delay can be estimated.

4.1 Time Transfer Basics

A half duplex two-way ranging method is described next [1, 2]. Figure 4.1 is a block diagram of the system whose transmission frames and timing are illustrated in Figure 4.2. This description is a simplified one since it makes the following assumptions:

- The time bases of both terminals have the same rate with no drift.
- Speed of propagation is the same in both directions.
- No delays other than that due to wave propagation in space.
- No multipath interference.

A and B are transceivers, each with an independent clock. The object is to find the time difference between the two clocks at an instant in time,

$$\Delta t = t_B - t_A \tag{4.1}$$

and the time of flight, or propagation time, between the terminals, T_p. The distance between terminals is $T_p c$, where c is the speed of light. The packet format may conform to a standard network protocol. It has a preamble followed by the message. At the end of the preamble the receiver uses a frame deliminator to determine the instant of the start of the message. This instant is the ranging reference point. The ranging initiator, terminal A, transmits a packet 1 and reads its clock, $t_{A,T}$, at the end of the preamble/beginning of message. The responder, terminal B, receives

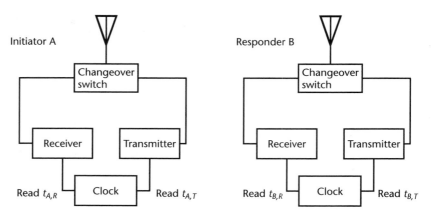

Figure 4.1 Simplified two-way ranging block diagram.

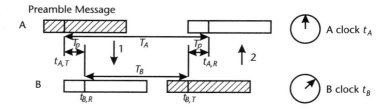

Figure 4.2 Two-way ranging timing.

the packet and reads its clock, $t_{B,R}$, when it detects the delimiter. After a period which includes the time to receive the whole packet plus a receive/transmit change-over, the responder B sends an acknowledge or response packet 2, and reads its clock, $t_{B,T}$, at the instant of the deliminator. Terminal A reads its clock, $t_{A,R}$, at the deliminator in the response packet it receives from B.

Using (4.1), two equations with two unknowns can be written:

$$\Delta t = t_{B,R} - T_p - t_{A,T} \tag{4.2}$$

$$\Delta t = t_{B,T} + T_p - t_{A,R} \tag{4.3}$$

Adding (4.2) and (4.3) gives the clock difference:

$$\Delta t = \frac{(t_{B,R} - t_{A,T}) - (t_{A,R} - t_{B,T})}{2} \tag{4.4}$$

The propagation time is found by subtracting (4.3) from (4.2):

$$T_p = \frac{(t_{B,R} - t_{A,T}) + (t_{A,R} - t_{B,T})}{2} \tag{4.5}$$

In order for the initiator to calculate Δt and t_p, the responder must send its clock readings $t_{B,R}$ and $t_{B,T}$ to the initiator, either in the response message or in a subsequent message.

The implementation of the ranging system in Figure 4.1 can be modified by substituting counters for blocks "clock." The transmission of the deliminator starts the counter at zero and the received deliminator stops it. The counter values when stopped are $t_{B,T}$ and $t_{A,R}$ and the propagation time, derived at A when it receives $t_{B,T}$ from B is

$$T_p = \frac{t_{A,R} - t_{B,T}}{2} \tag{4.6}$$

It is apparent that T_p depends on the durations

$$T_A = t_{A,R} - t_{A,T} \tag{4.7}$$

and

$$T_B = t_{B,T} - t_{B,R} \tag{4.8}$$

Then (4.5) and (4.6) can be expressed as

$$T_p = \frac{T_A - T_B}{2} \tag{4.9}$$

The resolution of the clock readings depends on the resolution of the time base of the transceivers. For distance resolution down to 1m, a system clock of better than 150 Mbps is required when the propagation delay is taken from one measurement. This could detect a round-trip time-of-flight increment of 6.67 ns, which corresponds to a two-way distance of 2m and a one-way distance of 1m. In order to achieve this resolution in a single measurement, the signal bandwidth must be at least equal to the sampling clock frequency. Sampling frequencies greater than the bandwidth improve resolution for multiple measurements that are reduced to a distance estimation by averaging.

4.2 Calibration Constants

In the simplified explanation above, it was assumed that the transmitted event, or epoch, was propagated over the distance to the receiver at the instant of the transmitter clock reading and that the receiver clock reading was taken at the instant of arrival. However, there are additional signal delays in the transmitter and receiver between measurements that must be accounted for in the time-of-flight calculations. These delays may be due for example to the group delays of RF analog filters and to circuit delays between correlator peak detection and reading of the value of the receiver or transmitter ranging counter.

Let $Tc_{x,T}$ equal the transmitter delay constant and $Tc_{x,R}$ equal the receiver delay constant as shown in Figure 4.3. x equals A or B. In the transmitter $Tc_{x,T}$ is added to the clock reading to express the instant that the signal reference point

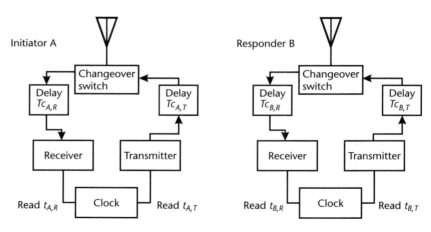

Figure 4.3 Ranging system diagram showing calibration delays.

leaves the antenna, and in the receiver $Tc_{x,R}$ is subtracted from the clock reading to the time that the reference point arrives at the antenna. Each terminal has its own set of constants: $Tc_{A,T}$ and $Tc_{A,R}$, $Tc_{B,T}$ and $Tc_{B,R}$. The constants are in units of clock cycles. Equation (4.5) is then modified as follows:

$$T_p = \frac{(t_{B,R} - Tc_{B,R} - t_{A,T} - Tc_{A,T}) + (t_{A,R} - Tc_{A,R} - t_{B,T} - Tc_{B,T})}{2} \qquad (4.10)$$

$$T_p = \frac{(t_{A,R} - t_{A,T}) - (Tc_{A,R} + Tc_{A,T}) - (t_{B,T} - t_{B,R}) - (Tc_{B,T} + Tc_{B,R})}{2}$$

One way to determine the calibration constants is by using a loopback connection. The transceiver is configured to receive its own transmission. The ranging counter is read at the reference point of the transmitted packet and read again when detected by the receive chain. The difference of the two readings is the sum of $Tc_{x,R}$ and $Tc_{x,T}$. Shielding must be used to be sure the transmitted signal enters the receiver chain at the antenna terminal.

The constants may be found separately by performing one-way ranging over a known distance with a second terminal whose constants are known. The clocks of the two terminals must be synchronized.

Normally the calibration constants are used only by the physical layers to which they refer. The times that are transmitted from one terminal to another for processing are corrected before transmission by adding or subtracting the calibration constants as appropriate.

4.3 Range Uncertainty

There are many real-life realities that make the ideal system described earlier unattainable although approachable. These are factors that should be taken into account for time transfer ranging over relatively short distances:

- Difference of time between the counter event and the time the deliminator instance leaves or enters the transceiver antenna (Section 4.2);
- Absolute and relative drifts of time bases of terminals;
- Pulse time resolution due to finite bandwidth and noise;
- Multipath interference.

4.3.1   Clock Drift and Measurement Time

Time base drifts reduce the accuracy of half duplex two-way ranging [3]. Assume that the nominal rate of the clocks of Figure 4.2 is R_0 ticks per second. The actual clock rates are

$$R_A = R_0(1 + \Delta) \tag{4.11}$$

$$R_B = R_0(1 + \Delta + \delta) \tag{4.12}$$

where Δ is the absolute drift of the clock at A from perfect time and δ is the relative drift of B's clock compared to that of A. Δ and δ can be positive or negative. In this discussion we do not consider the influence of the granularity of the clock counters.

Now the elapsed time readings of the A and B clocks are estimates that are expressed as follows:

$$T_A' = T_A(1 + \Delta) \tag{4.13}$$

$$T_B' = T_B(1 + \Delta + \delta) \tag{4.14}$$

where T_A and T_B are the true elapsed times of (4.7) and (4.8) and shown on Figure 4.2.

Referring to (4.9), the estimated propagation delay can then be expressed as

$$T_p' = \frac{T_A' - T_B'}{2}$$

$$T_p' = T_p + \Delta \cdot T_p - \frac{\delta \cdot T_B}{2} \tag{4.15}$$

Clock accuracy in a communication system will generally not be worse than around 20 ppm $= 2 \times 10^{-5}$. At this or better accuracy the second term on the right of (4.15) can be ignored, leaving

$$T_p' = T_p - \frac{\delta \cdot T_B}{2} \tag{4.16}$$

Equation (4.16) indicates that the propagation time estimation error, $\delta T_B/2$, is proportional to the difference in deviation of the two terminal clocks, that is, the clock drift between them, and the delay time of the responder's range acknowledgement. For a worst-case relative drift of 40 ppm, range errors for three values of T_B are shown in Table 4.1.

Table 4.1 Range Error Versus Responder
Elapsed Time for Interrogator-Responder Drift
Equal to 40 ppm

T_B (microseconds)	Range Error (meters)
10	0.12
100	1.2
1,000	12

There are three ways to reduce the effect of relative time base drift between interrogator and responder. A short ranging packet length is necessary for minimum error, as shown in Table 4.1. While message length can usually be kept short for ranging, the preamble length will be fixed by the protocol and will probably be the limiting factor determining the length of the ranging packet. Synchronizing the responder clock rate to the incoming packet during the preamble will also reduce the error.

A third way is to essentially cancel out the error term by performing back-to-back ranging, where range is measured with A being the interrogator and B the responder, then measured again with reversed roles: B is the interrogator and A the responder. Figure 4.4 shows the exchange of packets. Averaging the two measurements gives the final range estimate. The reduction of the error term is shown as follows. The result of the first range estimation, with A the interrogator and B the responder, adapted from (4.15), is

$$T'_{p,1} = T_p + \Delta \cdot T_p - \frac{\delta \cdot T_{B,1}}{2} \qquad (4.17)$$

For the second estimation B is the interrogator and A the responder, giving

$$T'_{p,2} = T_p + \Delta \cdot T_p + \frac{\delta \cdot T_{B,2}}{2} \qquad (4.18)$$

We now calculate a new estimate for the propagation time, T''_p, by averaging $T'_{p,1}$ and $T'_{p,2}$, getting:

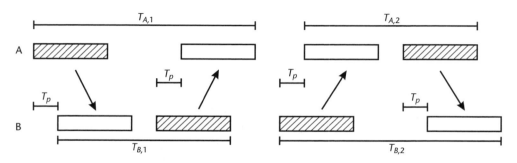

Figure 4.4 Back-to-back ranging timing diagram.

$$T_p'' = \frac{T_{p,1}' + T_{p,2}'}{2}$$

$$T_p'' = T_p + \Delta \cdot T_p + \frac{\delta}{4} \cdot (T_{B,2} - T_{B,1}) \qquad (4.19)$$

Comparing (4.19) and (4.15), only the last term has changed. When packet lengths and receive to transmit delay are the same for the two propagation time estimation trials, it can be seen from Figure 4.4 that $(T_{B,2} - T_{B,1})$ approximately equals $2T_p$. Substituting in (4.19) gives:

$$T_p'' \approx T_p + \Delta \cdot T_p + \frac{\delta}{2} \cdot T_p$$

$$T_p'' \approx T_p \left(1 + \Delta + \frac{\delta}{2}\right) \qquad (4.20)$$

The factor in parenthesis is very close to unity, so the result of the back-to-back propagation time estimation is to substantially reduce the measurement error due to the clock rate drift.

4.3.2   Noise

As explained in Chapter 2, noise and limited bandwidth reduce time instant resolution. In the time transfer method of distance measuring, noise can prevent an accurate determination of the reference point in the received signal, and thus an erroneous reading of the clock (Figures 4.1 and 4.2). The method requires at least two clock readings of the received reference point, and an error in one or both of them will affect the propagation time, and hence the distance, estimation. A high noise or interference level may disrupt the measurement all together, and the system should be designed to determine that a clock reading is not plausible and the measurement must be abandoned or tried again.

The degree to which noise affects the clock readings and the accuracy of the distance measurement is a function of signal-to-noise ratio, bandwidth, and the clock resolution. Noise causes jitter in the clock reading. While the clock itself is stable, the noise in the input signal produces a relative jitter of the sampling point. The relationship between the rms jitter σ_τ, signal bandwidth f_0, and the signal-to-noise ratio S/N can be approximated by [4]:

$$\sigma_\tau = \frac{1}{2 \cdot \pi \cdot f_0 \cdot (S/N)} \qquad (4.21)$$

Averaging of multiple measurements can improve accuracy at the expense of measuring time. By making N independent measurements, the improvement of the rms jitter is:

$$\sigma_{av} = \frac{\sigma_\tau}{\sqrt{N}} \qquad (4.22)$$

4.3.3 Multipath

Multipath interference creates the possibility that the propagation delay on a path other than the direct path between transmitter and receiver will be measured erroneously as time of flight for the purpose of finding the range. For example, if the deliminator is detected by a matched filter, multipath signals will result in multiple output pulses that indicate the timing epoch. The receiver should be capable of examining the pulse times and deciding which is the desired one. The earliest pulse of those that arrive within the expected range of multipath reflections should represent the direct path, but it may very well not be the strongest pulse. The direct path may be blocked and not detected at all, in which case the estimated range will be too high. When a correlator is used to detect the deliminator, a chip period that is short compared with the multipath span facilitates rejection of multipath reflections because of the autocorrelation properties of the deliminator. For indoor systems particularly, path lengths are short and a relatively high chip rate must be implemented. This makes UWB (ultrawideband) systems inherently advantageous for distance measuring (see Chapter 11).

4.3.4 Relative Motion

Relative motion between the interrogator and the responder could affect the distance measurements. In a single one-way time transfer measurement, relative motion will effect the propagation time estimate if the distance between the terminals changes more than the wavelength of the measuring clock frequency during the measurement. The bound on the relative velocity that will not affect the distance measurement is [6]:

$$v < \frac{T_C \cdot c}{T_M} \tag{4.23}$$

where v is the relative velocity, T_C is the period of the clock frequency, T_M is the total measurement time, and c is the speed of light. An example shows the degree to which target velocity is apt to affect the distance measurement in a short-range environment.

Example 4.1

In an ultrawideband distance measuring system we assume the following parameters:

Clock rate, $f_C = 528$ MHz
Measurement time, $T_M = 50$ microseconds
$c = 3 \times 10^8$ m/s

$$v_{max} = \frac{(1/f_C) \cdot c}{T_M} \tag{4.24}$$

$v_{max} = 11.364$ km/s

It is clear that during a single measurement, the relative velocity is not a factor. However, if many measurements are taken over a much longer period of time in order to improve accuracy with noisy signals, v_{max} is reduced proportionally to the increase in the measurement time.

4.4 Ranging Procedure in Wireless Network

An example of distance measurement capability and measurement procedure is provided by a standard for an ultrawideband WPAN (wireless personal area network), ECMA-368, developed by Ecma International, an industry association dedicated to the standardization of information and communication technology and consumer electronics [5]. The characteristics of the specified network terminals that relate to distance measurement are the following:

- Ranging clock rate: 528 MHz and optional rates of 1,056 MHz, 2,112 MHz, and 4,224 MHz;
- Clock accuracy: 20 ppm;
- Ranging accuracy: 60 cm or better;
- Ranging reference point: In preamble, at end of frame synchronization sequence at first sample of channel estimation sequence;
- Ranging calibration constants:
  1. Ranging transmit delay—time from sampling the outgoing ranging reference point to the time that the reference point is emitted from the antenna;
  2. Ranging receive delay—time from arrival of the reference point at the antenna to the time of its sampling by the ranging clock.
- Approximate measuring interval: 23.4 μs plus time of flight. Includes 10-μs preamble, 3.4-μs message frame, and 10-μs short interframe spacing.

A distance measuring procedure for ECMA-368 is illustrated in Figure 4.5. Device A is the interrogator and device B is the responder. The terminals are represented by a physical layer (PHY), medium access control (MAC), and in the case of A, a management entity. The details of the steps are as follows.

(1) Device A initiates the range measurement. The range request specifies the number of consecutive measurements to be performed, up to 256. The results of the measurements can be averaged to improve resolution and accuracy.

(2), (3) Device A MAC turns on the range timer and transmits a range measurement request command frame to device B, which includes the number of measurements requested.

(4), (5) Upon receiving the range measurement request, device B turns on its range timer and returns an acknowledgment of the received frame.

(6) Device A sends a range measurement frame to device B.

(7) Device A reads its range timer on transmission of the reference point and device B reads its range timer on reception of the reference point.

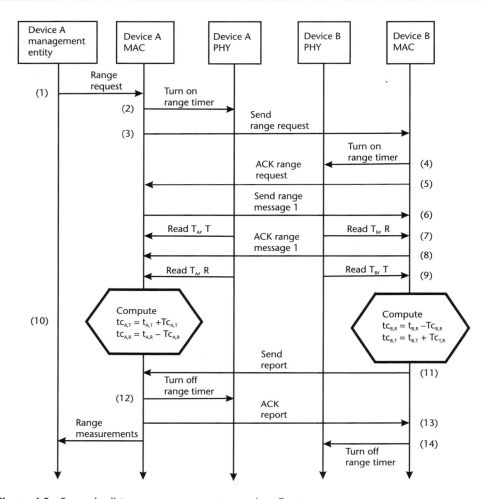

Figure 4.5 Example distance measurement procedure flow.

(8)   Device B returns an acknowledge frame to device A.

(9)   Device B reads its range timer on transmission of the reference point of the acknowledge frame and device A reads its timer on reception of the reference point.

(10) Devices A and B modify their timer reading with their individual calibration constants. The results are used in the range equation.

(11) Device B transmits to A a range measurement report containing the modified results of its timer readings. The number of sets of measurement results in the report corresponds to the number of measurements requested in step (3).

(12), (13) Device A turns off its range counter and sends an acknowledgment of receipt of device B's report.

(14) Device B turns off its range counter. Device A proceeds to calculate a single estimation or multiple estimations of the range. The expression for a single estimation is:

$$d = c \cdot \frac{(tc_{A,R} - tc_{A,T}) - (tc_{B,T} - tc_{B,R})}{2} \qquad (4.25)$$

where d is the range and c is the speed of light.

4.5 Conclusions

The time transfer method is a relatively simple way to make distance measurements between two terminals whose clocks are not synchronized. The time for an individual measurement is short but a clock period comparable to the distance resolution required must be available. For short-range, indoor ranging, wide bandwidth communication systems are used, in particular, UWB. In location-aware sensor networks, the time transfer method can also facilitate one way ranging. For example, periodic two-way time transfer between a sensor device and a controller beacon can maintain clock synchronization. During a period where clock drift does not exceed required distance resolution, the sensor can make one-way range estimates.

References

[1]   Batra, A., et al., Slide 164, "UWB Ranging Via Two-Way Time Transfer," IEEE 802.15-03/267r2, September 2003.

[2]   Hanson, D. W., "Fundamentals of Two-Way Time Transfer by Satellite," *43rd Annual Frequency Control Symposium*, 1989, pp. 174–178.

[3]   Benoit, D., "Ranging Protocols and Network Organization," IEEE P802.15-04/427r0, IEEE 802.15 Working Group for Wireless Personal Area Networks, August 2004.

[4]   Maxim Integrated Products, Application Note 800, "Design a Low Jitter Clock for High Speed Data Converters," November 20, 2001.

[5]   ECMA Standard ECMA-368, "High Rate Ultra Wideband PHY and MAC Standard, First Edition," December 2005.

CHAPTER 5

Multicarrier Phase Measurement

Up to now we have seen methods of distance measurement that depend on discrete time unit counting to determine time of flight of a radio signal between transmitter and receiver. Both the spread spectrum sequence alignment method and time transfer procedure achieve higher resolution of distance through greater clock rates and signal bandwidth. In both of them the measuring stick is the system clock and the graduations are time base pulses or fractions of them. In this chapter distance is measured by measuring phase differences of RF carriers and in general we deal in the frequency domain as compared to the time domain and the realm of pulse widths and rise times. Still the basic principles take hold—for example, higher resolution is gained by spectrum expansion. Multicarrier distance measuring techniques discussed here are applied using frequency hoping spread spectrum and OFDM transmissions.

5.1 Principle of Multicarrier Phase Measurement

The phase difference between a received CW carrier, or a modulated tone, and a reference signal, can be used to measure distance, as demonstrated in Figure 5.1. An interrogator transmits an uninterrupted tone. The responder phase locks a local oscillator to the incoming tone and retransmits it. The interrogator measures the phase lag of the received tone compared to the reference tone and calculates the one way distance as:

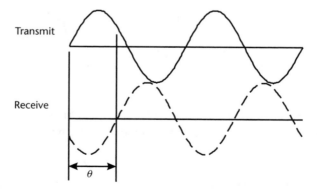

Figure 5.1 Single carrier distance measurement.

$$d = \frac{\lambda}{2} \cdot \left(\frac{\theta}{2\pi} + n \right) \qquad (5.1)$$

where λ is the tone wavelength, θ is the phase, and n is an integer. Since the range of the phase measurement is between 0 and 2π, it would be necessary to keep track of the number of whole cycles, n, that passed in order to determine one-way distances greater than $\lambda/2$. The ambiguity can be eliminated by sending two tones and measuring the difference between their received phases, each phase being compared to a reference. Using frequency instead of wavelength in (5.1), we have for each of the two tone frequencies

$$\theta_1 = 2\pi \cdot \left(\frac{2 \cdot d \cdot f_1}{c} - n \right) \qquad (5.2)$$

$$\theta_2 = 2\pi \cdot \left(\frac{2 \cdot d \cdot f_2}{c} - n \right) \qquad (5.3)$$

where c is the speed of light. Subtracting (5.2) from (5.3) and solving for d:

$$d = \frac{c}{4\pi} \cdot \frac{\theta_2 - \theta_1}{f_2 - f_1} \qquad (5.4)$$

Now the ambiguity in range has been eliminated. The span of the measurement of θ is 2π, so the maximum value of d that can be measured using two phase difference measurements is conditioned on a maximum difference between the two measurement frequencies. For example, if f_2 and f_1 differ by 1 MHz, the maximum measurable one-way distance, using (5.4), is 150m.

As $f_2 - f_1$ is made smaller to accommodate longer range, the resolution or range error increases. Let $\delta\theta$ equal a given phase difference measurement error. The distance error δd is:

$$d + \delta d = \frac{c}{4\pi} \cdot \frac{\Delta\theta + \delta\theta}{\Delta f} \qquad (5.5)$$

$$\delta d = \frac{c}{4\pi} \cdot \frac{\delta\theta}{\Delta f} \qquad (5.6)$$

where $\Delta\theta = \theta_2 - \theta_1$ and $\Delta f = f_2 - f_1$. It is clear from (5.6) that the error increases in inverse proportion to the measurement frequency difference.

5.2 Phase Slope Method

A system for measuring range by the phase difference method is shown in Figure 5.2. An interrogator transmitter transmits an unmodulated carrier on a frequency f_0. The responder, whose details will be given later, locks its local oscillator in

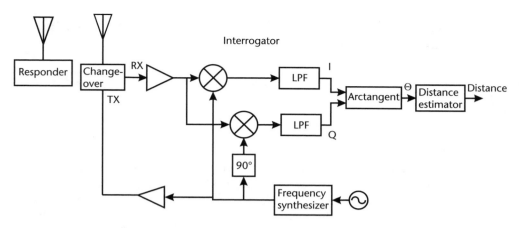

Figure 5.2 Phase measuring interrogator and responder.

frequency and phase to the received signal and retransmits it. The interrogator switches over to receive and measures and stores the phase difference between the received carrier and its local oscillator. The interrogator and responder then switch to a new frequency and the process repeats itself, with a new phase difference stored. Note that phase coherency between the original frequency signal and the second frequency is not required, but the responder must lock its synthesizer to the received signal and maintain continuous phase during changeover from receive to transmit.

The use of two frequencies to make phase difference measurements is relatively simple but not particularly practical in a wireless environment. We have seen that reducing frequency separation to accommodate larger range reduces accuracy. Also, one of the frequency channels could be occupied by an interferer, thus making the distance measurement impossible. A way to improve performance of the phase difference scheme is to adapt frequency hopping spread spectrum techniques [1]. Making phase difference measurements on more than two separated frequencies improves resolution and accuracy and also gives redundancy for the case where one or more frequency channels are occupied by another signal. This enhancement is well suited to the system of Figure 5.2, since the frequency synthesizer can be adjusted to a wide range of frequency channels.

Using the multifrequency approach, the phase and frequency differences can be plotted as consecutive adjacent points on a discrete phase versus frequency curve, derived from (5.2) or (5.3):

$$\theta_i = \frac{4\pi}{c} \cdot d \cdot f_i + C \qquad (5.7)$$

where f_i is the frequency and C, the intercept on the phase axis, is not necessarily constant over the whole range of f_i. i is an integer ranging from 0 to the number of frequencies less 1.

Each phase point θ_i is the phase difference between the reference output of the synthesizer in the interrogator of Figure 5.2 and the incoming signal from the responder. The one-way distance d is easily found from the slope of the curve created from (5.7), which is

$$slope = \frac{4\pi}{c} \cdot d \tag{5.8}$$

$$d = \frac{c}{4\pi} \cdot slope$$

The principle of the phase slope method of range measurement can be demonstrated by creating a set of data and then extracting the range as shown in the receiver of Figure 5.2. For an example, let's decide to use FHSS with 20 hopping frequencies separated by 1 MHz, extending from 905 to 924 MHz. To create the data for the example, we assume a one-way distance of 20m and calculate a table of phase angles between 0 and 2π using the expression

$$\theta_i = \frac{4\pi \cdot d \cdot f_i}{c} \mod (2\pi) \tag{5.9}$$

On the lowest frequency, $f_0 = 905$ MHz, the phase is $\theta_0 = 4.189$ radians = 240°. Similarly, the phase data θ_i for all hopping channels f_i can be found.

In order to make the example correspond to the operation shown in of Figure 5.2, I and Q values are derived as $I_i = \cos(\theta_i)$ and $Q_i = \sin(\theta_i)$. An arctangent function, shown as a block in Figure 5.2, computes the phase. The phase values at the output of the arctangent block, which are labeled Θ_i, range between $-\pi$ and $+\pi$. At $f_0 = 905$ MHz, for example, $\Theta_0 = \arctan[\sin(4.189)/\cos(4.189)] = -2.094 = -120°$.

A plot of frequency versus the phase measured by the arctangent block is shown in Figure 5.3. The curve has positive slope straight line sections that break when the arctangent calculation exceeds its limit of $\pm\pi$ radians. To make the process more meaningful and more realistic, random noise is included in the I, Q outputs and thus in the example phase measurements. This makes the plotted output of the arctangent block look like Figure 5.4. A simple algorithm can be applied to make the curve slope continue in a positive direction for the whole frequency

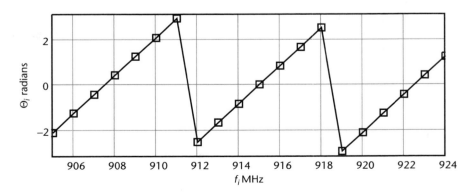

Figure 5.3 Phase versus frequency plot of FHSS signal.

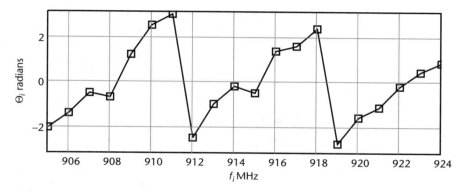

Figure 5.4   Phase versus frequency plot of FHSS with random noise.

span of the measurements. The algorithm is described as follows, where the new straightened phase values are labeled φ_i.

Set $\varphi_0 = 0$

For $i > 0$

If $(\Theta_i - \Theta_{i-1}) < -\epsilon$ (5.10)

set $\varphi_i = \varphi_{i-1} + (\Theta_i - \Theta_{i-1}) + 2\pi$

Else

set $\varphi_i = \varphi_{i-1} + (\Theta_i - \Theta_{i-1})$

The value of ϵ is between 0 and 2π radians. It should be very close to zero when the expected distance is close to the limit determined by the adjacent channel frequency separation, and larger for short distances in the presence of measurement noise.

The new values of φ after application of the algorithm are plotted in Figure 5.5. This is a scatter diagram with discrete frequencies along the abscissa and noisy phase points on the ordinate. The figure also contains a least mean square regression line whose slope gives an estimate of propagation delay that can be used to estimate range using (5.8). The estimate of the regression line is discussed in detail later in this chapter. It should be noted that the actual channel frequency does not enter into the determination of range—only the increment between channels. Remember that the above development is based on the one-way distance between two terminals engaged in a back and forth communication. The total propagation distance is twice that given in (5.8).

5.3   Phase Error Versus Signal-to-Noise Ratio

The relationship between phase error and signal-to-noise ratio depends on the difference between adjacent hopping frequencies and the number of hopping channels. The aim is to determine those parameters that will give a required range

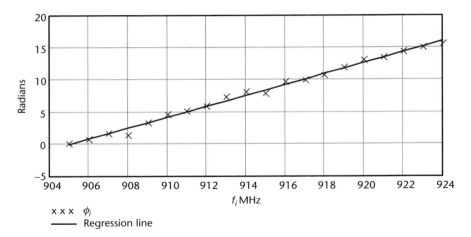

Figure 5.5 A straightened phase versus frequency plot with data and linear regression line.

accuracy at a minimum signal to noise ratio. First, we find the statistics of the phase estimate for an individual channel measurement [2]. Figure 5.6 shows a quadrature phase detector whose input is a CW signal of angular frequency ω_0 plus additive white Gaussian noise (AWGN) whose one sided spectral density equals N_0. The input signal is:

$$r(t, \theta) = A \cdot \cos(\omega_0 t + \theta) + n(t) \qquad (5.11)$$

The phase of the input signal, θ, is to be compared with a local oscillator output whose quadrature components $2\cos(\omega t)$ and $-2\sin(\omega t)$ are applied to the mixer of the phase detector. The factor of 2 and the negative sign of the quadrature component are chosen for convenience but do not affect the final result.

The narrowband random noise is expressed as follows [3]:

$$n(t) = n_c \cdot \cos(\omega_0 t) - n_s \cdot \sin(\omega_0 t) \qquad (5.12)$$

The quadrature envelopes n_c and n_s have zero mean and variance:

$$\sigma_n^2 = B_n \cdot N_0 \qquad (5.13)$$

where B_n is the noise bandwidth of the lowpass filters (LPF) in Figure 5.6.

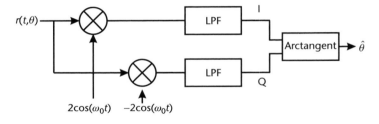

Figure 5.6 Phase detector.

After performing multiplication in the mixers, double frequency components are blocked by the lowpass filters whose outputs are

$$I(\theta) = A \cdot \cos(\theta) + n_c \tag{5.14}$$

$$Q(\theta) = A \cdot \sin(\theta) + n_s \tag{5.15}$$

$I(\theta)$ and $Q(\theta)$ are Gaussian random variables with mean values $A\cos(\theta)$ and $A\sin(\theta)$, and variance σ_n^2. $I(\theta)$ and $Q(\theta)$ are statistically independent, so their joint probability density is the product of their individual probability densities and can be written as

$$p(I, Q) = \frac{1}{2\pi \cdot \sigma_n^2} \cdot \exp\left(\frac{-[(I - A \cdot \cos(\theta))^2 + (Q - A \cdot \sin(\theta))^2]}{2 \cdot \sigma_n^2}\right) \tag{5.16}$$

An estimate of the phase, θ_e, is taken from the output of the arctangent block in Figure 5.6.

$$\theta_e = \arctan\left(\frac{Q(\theta)}{I(\theta)}\right) \tag{5.17}$$

θ_e is a random variable with mean value θ, and variance σ_θ^2 which is to be determined.

In order to find the variance of the estimated phase, we need to know its probability density function, $p_\theta(\theta_e)$. On the way, a joint density function, $p(V, \theta_e)$ is found, from a transformation of variables in (5.16). V^2 is a measure of the power at the output of the lowpass filters:

$$V^2 = I^2 + Q^2 \tag{5.18}$$

Observing (5.16), it is clear that the terms containing θ are means that shift the position of the probability density function but do not affect the variance. So, to simplify the probability functions that follow, θ is set to zero. Performing the transformation of variables [3] on (5.16) using (5.17) and (5.18) we get

$$p(V, \theta_e) = \frac{V}{2\pi\sigma^2} \cdot \exp\left(\frac{-[V^2 + A^2 - 2 \cdot A \cdot V \cdot \cos(\theta_e)]}{2 \cdot \sigma^2}\right) \tag{5.19}$$

The result that we are looking for is the variance of the phase estimate as a function of the signal-to-noise ratio at the input to the phase detector, which is

$$SNR = \frac{A^2}{2 \cdot \sigma_n^2} \tag{5.20}$$

It is the SNR that is of interest, not the actual values of A and σ_n^2, so σ_n^2 can be set to 1 and A in (5.19) substituted by $A = \sqrt{2 \cdot SNR}$.

Now to derive the density function of θ_e from the joint density function in (5.19), the latter is integrated over the range of V:

$$p_\theta(\theta_e, SNR) = \int_0^\infty p(V, \theta_e)\, dV \qquad (5.21)$$

$$p_\theta(\theta_e, SNR) = \int_0^\infty \frac{V}{2\pi} \cdot \exp\left(\frac{-[V^2 + 2 \cdot SNR - 2 \cdot \sqrt{2 \cdot SNR} \cdot V \cdot \cos(\theta_e)]}{2}\right) dV$$

No attempt is made here to simplify (5.21), but the phase angle probability density function can be used in its form as a definite integral by a mathematical program such as MATHCAD [4]. Then the variance of the phase estimate can be calculated using

$$\mathrm{var}_\theta(SNR) = \int_{-\pi}^{\pi} \theta_e^2 \cdot p_\theta(\theta_e, SNR)\, d\theta_e \qquad (5.22)$$

remembering that the mean of θ_e, θ, was set to zero.

The probability density of the estimated value of the phase for signal-to-noise ratio parameters of 4 (6 dB), 8 (9 dB), and 12 (11 dB), with mean phase = 0, is plotted in Figure 5.7.

Table 5.1 shows phase variance versus SNR over a range from 2 to 20 dB. It is evident that an approximation for the phase variance for large SNR (greater than around 8 dB) is

$$\mathrm{var}_\theta(SNR) \approx \frac{1}{2 \cdot SNR} \qquad (5.23)$$

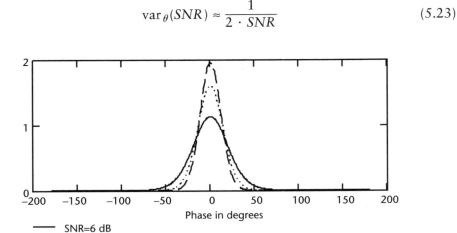

Figure 5.7 Probability density of the phase with the signal-to-noise ratio as parameter.

Table 5.1   Phase Variance Versus SNR

SNR_{dB}	SNR	Phase Variance (rad^2)
2	1.6	0.48
4	2.5	0.28
6	4.0	0.15
8	6.3	0.088
10	10	0.053
16	39.8	0.013
20	100	0.005

5.4   Estimation of Distance Variance Versus SNR

In this section the results of the phase variance versus SNR are used to determine the accuracy of the distance measurement as a function of the SNR, the number of hopping channels, and the separation between them. The total distance is directly proportional to the propagation delay, which can be estimated by $1/2\pi$ times the slope of a linear mean square regression line that is drawn through the phase versus frequency curve data points, to which the algorithm of (5.10) has been applied. Figure 5.5 shows an example of the regression line. Only the slope of the line is relevant to finding the propagation time and distance, not the intercept points on the axes.

The least mean square slope of the regression line for the plot of frequency versus phase is found according to [3]

$$slope = \frac{E[(f - f_m) \cdot (\theta - \theta_m)]}{E[(f - f_m)^2]} \qquad (5.24)$$

where $E(\ )$ is the expectation, f is the frequency variable, and θ is the value of the phase measurement. The subscripted m values are means:

$$f_m = \frac{1}{N} \sum_{i=0}^{N-1} f_i \qquad (5.25)$$

$$\theta_m = \frac{1}{N} \sum_{i=0}^{N-1} \theta_i \qquad (5.26)$$

where N is the number of channels.

Expression (5.24) can also be expressed as

$$slope = \frac{E(f \cdot \theta) - f_m \cdot \theta_m}{E(f^2) - f_m^2} \qquad (5.27)$$

The expectations are written in the following expression as discrete averages, where i is the consecutive hopping channel subscript and N is the number of channels:

$$slope = \frac{\frac{1}{N} \cdot \sum_i (f_i \cdot \theta_i) - f_m \cdot \theta_m}{\frac{1}{N} \cdot \sum_i f_i^2 - f_m^2} \tag{5.28}$$

The average slope is needed in order to find the slope's variance:

$$slope_{av} = E(slope) \tag{5.29}$$

$$slope_{av} = \frac{\frac{1}{N} \cdot \sum_i (f_i \cdot \theta_i) - f_m \cdot \theta_m}{\frac{1}{N} \cdot \sum_i f_i^2 - f_m^2}$$

The variance of the slope is

$$\mathrm{var}_{slope} = E(slope\text{-}slope_{av})^2 \tag{5.30}$$

After making the evident substitutions, and using the fact that the phase readings on the different channel frequencies are uncorrelated, the variance of the slope is found to be

$$\mathrm{var}_{slope} = \frac{\frac{1}{N^2} \cdot \mathrm{var}_\theta \cdot \sum_i f_i^2}{\left[\frac{1}{N} \cdot \sum_i f_i^2 - \left(\frac{1}{N} \cdot \sum_i f_i\right)^2\right]^2} \tag{5.31}$$

The value of var_θ can be estimated from measured phase data using

$$\mathrm{var}_\theta = \frac{1}{N} \cdot \sum_{i=0}^{N-1} (\theta_i - \theta_m)^2 \tag{5.32}$$

The object is to find the variance of the slope as a function of N and of the separation between hop frequencies, Δf. The set of hop channel frequencies can be translated down to start at 0 Hz, without affecting the slope:

$$f_i = \Delta f \cdot i \text{ for } i = 0 \ldots N - 1 \tag{5.33}$$

After substituting in (5.31), the result is

$$\mathrm{var}_{slope}(\Delta f, N, \mathrm{var}_\theta) = \frac{\mathrm{var}_\theta \cdot \sum_{i=0}^{N-1} i^2}{\Delta f^2 \cdot \left[\sum_{i=0}^{N-1} i^2 - \frac{1}{N} \cdot \left(\sum_{i=0}^{N-1} i\right)^2\right]^2} \tag{5.34}$$

The variance of the one-way distance is expressed using (5.8):

$$\text{var}_d(\Delta f, N, \text{var}_\theta) = \left(\frac{c}{4\pi}\right)^2 \cdot \text{var}_{slope}(\Delta f, N, \text{var}_\theta) \tag{5.35}$$

Figures 5.8 and 5.9 are plots of the standard deviation of the one-way distance versus number of measurements, N, and channel spacing, Δf, over a range of parameter values. In both figures, the phase variance is 0.088, corresponding to a signal to noise ratio of 8 dB. The figures indicate how distance accuracy improves when increasing the number of hopping channels and channel frequency separation.

The variance of the phase measurement, var_θ, is dependent on the signal-to-noise ratio SNR [see (5.22) and (5.23)], so the variance of the distance, (5.35), is indirectly dependent on the SNR. The Chebyshev inequality, (5.36), can assist in assessing the distance accuracy that can be obtained from a set of the three parameters Δf, N, and SNR.

$$\text{Prob}\left(\left|d_e - d\right| \geq \epsilon\right) \leq \frac{\text{var}_d}{\epsilon^2} \tag{5.36}$$

The inequality sets an upper bound to the probability that the distance error exceeds a given value ϵ. From (5.23) or Table 5.1 we find the phase variance for a given signal-to-noise ratio. For SNR = 8 dB the phase variance is 0.088. Assume the distance measuring system uses 20 hop channels with channel separation equal to 1 MHz. From Figure 5.8 the distance variance is found to be $(0.53\text{m})^2 = 0.28 \text{ m}^2$. For a desired accuracy of 1m, applying (5.36) indicates that the probability of exceeding that value is upper bounded by 28%.

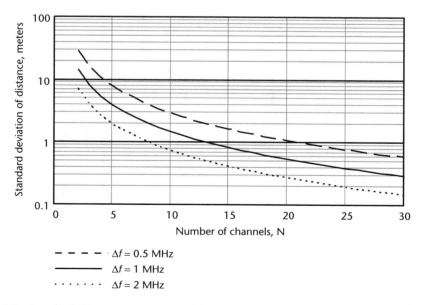

Figure 5.8   Standard deviation of one-way distance versus number of channels, N, for 8-dB SNR, for delta f = 0.5 MHz, 1 MHz, and 2 MHz.

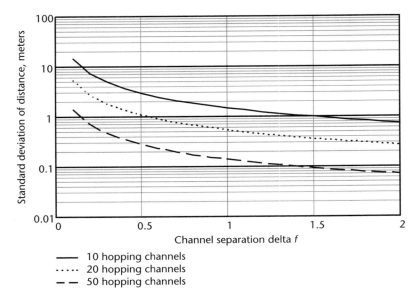

Figure 5.9 Standard deviation of distance versus channel separation, Δf, for 8-dB SNR and the number of hopping channels equals 10, 20, and 50.

5.5 Multipath

Multipath propagation is a particularly detrimental impediment to a wireless distance measuring system. The desired range is the line-of-sight distance between two wireless terminals, but propagation over other paths takes more time and therefore could be interpreted as longer distance. In the case of the phase slope method of distance measurement, multipath propagation results in more than one slope, and the slope of the regression line will not give the correct result.

In most wireless applications, means may be taken to reduce multipath interference, but it does not matter if these measures affect the propagation time of the signal, since the aim is to improve the signal-to-noise ratio or bit error rate. However, in the case of a distance measuring system, it is just the propagation time that is desired. Solutions that have been developed to deal with multipath for wireless communications are not necessarily applicable for distance measurement.

Figure 5.10 is a schematic representation of the multipath phenomenon. The transmitted signal reaches the receiver over three different path lengths, giving propagation times of t_1, t_2, and t_3. The strengths of arriving signals over the three paths are affected by the path length, the nature of the reflection or diffraction on the echo signals, and attenuation through different media (obstructions). These different signal strengths are represented by A_1, A_2, and A_3.

Three types of interference can result from multipath:

1. The signal strength fluctuates rapidly over relatively small distance changes between the transmitter or receiver, or over small time periods when the reflectors move in relation to the communicating terminals.

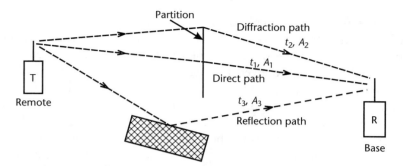

Figure 5.10 The multipath phenomenon.

2. Time dispersion, or echoes, spread the transmitted pulses in the case of digital communication and thereby cause intersymbol interference. In the case of analog television, this interference type results in "ghosts" of the picture details.
3. Different Doppler shift frequencies are created due to different degrees of changes in path length when the communicating terminals are in relative motion, one in respect to the other.

Common methods employed to combat multipath interference include:

1. Diversity reception. In this method, two or more signals with relatively noncoherent amplitudes are produced through space (separated antennas), frequency, time, or antenna polarity diversity. The strongest between the diverse samples is selected for detection or demodulation.
2. Use of an adaptive equalizer. This method, applicable only to digital modulated signals, passes the received signal through a tapped delay line where adaptively adjusted tap take off parameters can cancel out the echoes.
3. Rake receiver. A direct sequence spread spectrum receiver applies several time-delayed versions of the known transmitter pseudorandom sequence to decorrelate the incoming signal. Thus, the signal from the direct path (if there is one) and the strongest echoes may be decorrelated and combined to give a lower bit error rate than that obtainable from a signal over any one of the paths. This method is an implementation of time diversity reception.
4. Use of directional antennas at one or both terminals.

Of these methods, only number 4 could be used for distance measuring systems since it acts to strengthen the direct path signal in relation to the echoes. However, in applications where device orientation is uncontrollable, as in virtually all potential Bluetooth product uses and most other short-range links and networks, the antennas will have to be omnidirectional.

In the following we look at a method for analyzing distance measuring returns using Fourier transforms, which holds promise for improving the accuracy of the phase slope method of distance measurement in a multipath environment.

The phase slope method for measurement of propagation delay is based on the principle that the phase angle of the incoming signal is a linear function of frequency. The slope of the frequency versus phase delay curve (a straight line) obtained from the received frequency hopping signal is directly proportional to the propagation delay and thus to the distance between transmitter and receiver. Echoes that travel over longer paths than the direct path signal combine vectorally with that signal and cause the receiver to measure a greater distance than exists over the line-of-sight path. The echoes distort the linear phase versus frequency curve, and this distortion can give a clue as to the existence of multiple paths. However, we prefer to find another "dimension" for additional information that will allow us to correct the distance estimation.

The additional information we need can be found in the amplitude versus frequency profile of the received signal. When direct path and echo signals combine vectorally, their resultant amplitude, A which is a function of frequency f, for two signals, is

$$A(f) = \sqrt{a_1^2 + a_2^2 + 2 \cdot a_1 \cdot a_2 \cdot \cos \alpha(f)} \qquad (5.37)$$

where a_1 and a_2 are the amplitudes of the direct path signal and the echo signal and α is their phase difference. The phase difference here is a function of frequency and its value in radians equals the difference in distance traveled between the direct path and the echo path, divided by the wavelength and multiplied by 2π.

The effect of multipath propagation on the distance estimation obtained from the phase slope is shown in Figure 5.11. The solid line curve was produced by drawing the straightened phase versus frequency data of the sum of three complex envelope (baseband) signals. The composite baseband signal is

$$S_i = A_0 \cdot e^{j2\pi\tau_0 f_i} + A_1 \cdot e^{j2\pi\tau_1 f_i} + A_2 \cdot e^{j2\pi\tau_2 f_i} \qquad (5.38)$$

The times of flight of the direct signal and the two echoes are τ_0, τ_1, and τ_2 and corresponding amplitudes are A_0, A_1, and A_2. f_i are the hopping frequencies. The data for the curve is:

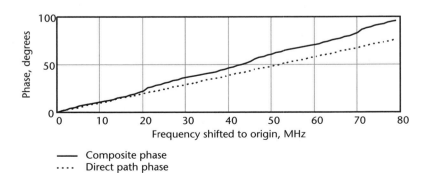

Figure 5.11   Phase versus frequency for direct path and composite multipath signals. The frequencies are shifted to start at origin.

$$A_0 = 1, \ \tau_0 = 153 \text{ ns}; \ A_1 = 0.5, \ \tau_1 = 190 \text{ ns}; \ A_2 = 0.7, \ \tau_2 = 235 \text{ ns}$$

There are 80 hopping channels from 2,400 MHz with 1-MHz increments. The slope of the regression line gives a time of flight of 193 ns, a considerable error from the time of flight of the direct wave, which is 153 ns. The phase versus frequency plot of the direct component alone compared to the multipath signal phase versus frequency is shown in Figure 5.11.

In order to find the true time of flight it is necessary to separate the indirect path signals from the direct path signal and to do this additional information is required. This can be obtained from a plot of the composite signal amplitude versus frequency, shown in Figure 5.12. Due to vectorial summing of the multipath signals at the receiver on each hop frequency, the signal amplitude is a function of frequency. The information contained in the amplitude versus frequency data can be extracted by a Fourier transform. The data used in the Fourier analysis are the I and Q voltage values in the phase detector (Figure 5.6). The arctangent block is not needed. Taking the complex direct Fourier transform of the frequency versus phase data separates the direct signal from the multipath returns in a plot of relative amplitude versus propagation time. This is in contrast to the usual use of the direct transform to go from the time to the frequency domain. The phase straightening algorithm is not used for this action. Zero stuffing can be used to make the number of samples used in the transform a power of 2, and to increase the resolution of the result in the time domain. Figure 5.13 shows the Fourier transform based on 80 data points, corresponding to the 80 hop channels, with no zero stuffing. Time resolution is 12.5 ns. The direct line of sight is identified as being the earliest peak. In order to prevent leakage artifacts from being considered as a legitimate signal, a threshold should be used for determining true echoes. Choice of the threshold value is a compromise between false alarm probability—detection of an echo where it does not exist, and the probability of missing a weak line-of-sight return. In Figure 5.13, the first echo (the line-of-sight return) is measured at a propagation time of 150 ns, whereas the actual time, used for creating the data, is 153 ns,

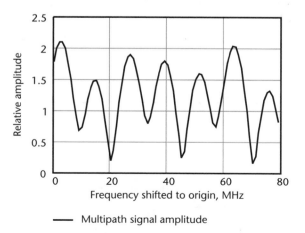

—— Multipath signal amplitude

Figure 5.12   Multipath composite signal amplitude versus frequency. The frequencies are shifted to start at origin.

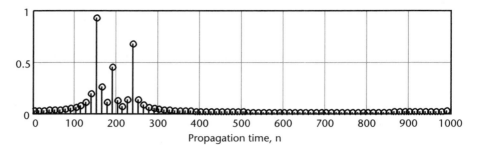

Figure 5.13   Fourier transform of multipath signal with 80 samples.

giving an error of one meter. In the plot of Figure 5.14 the Fourier transform is calculated for 1,024 samples, where zeros are used for all samples over 80. The resolution is now 0.977 ns. The maximum value on the first peak is at 153.32 ns, equivalent to one-tenth of a meter over the true value.

The basic system parameters for the Fourier transform analysis are the channel increments, Δf, and the total frequency span, $N\Delta f$, where N is the number of hops. The maximum propagation delay that can be displayed is

$$t_{\max} = 1/\Delta f$$

and the accuracy and ability to separate close echoes is a function of the frequency span. As stated, the resolution depends on the number of samples used to take the Fourier transform, but the accuracy suffers when the number of measured samples is low. In order not to affect the accuracy of the result, the inverse of the frequency span should be less than the difference of arrival time of the echoes. This is the same conclusion arrived at for direct TOA measurement methods.

In this discussion of multipath, noise was not considered, although in reality it is always present and will degrade results. The results of the proceeding section relating to how the SNR, number of samples and channel separation affect the estimated propagation delay accuracy apply to the use of Fourier transforms as well.

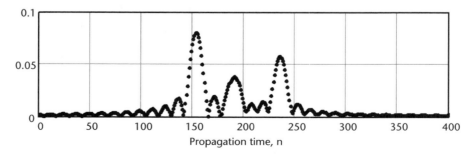

Figure 5.14   Fourier transform of multipath signal with 1,024 samples.

5.6   System Implementation

In one implementation of a frequency hopping two-way distance measuring system, the CW signals sent by an interrogator are phase locked by the responder and then transmitted back to the interrogator. Figure 5.15 is an example of an implementation method using analog phase lock loops. In the interrogator, a crystal oscillator serves as a common reference for three frequency synthesizers, labeled FS1, FS2, FS3. Dividers on the internal VCO outputs and reference inputs are contained in the FS blocks. The crystal oscillator frequency must be divided down to the channel hop frequency difference, or a submultiple. FS1 provides the output transmitted frequency, FS2 the receiver local oscillator frequency, and FS3 converts the IF amplification and filter chain output during receive to quadrature baseband signals that are used by the processor to compute range, as described in Section 5.2.

The responder's role is to phase lock on to the incoming signal from the interrogator, and maintain phase coherence after the received signal has ended and

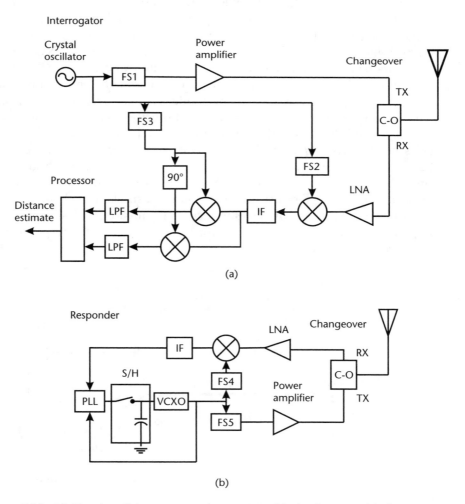

(a)

(b)

Figure 5.15 Multicarrier distance measuring system block diagram: (a) interrogator, and (b) responder.

during transmission of the reply. A voltage controlled crystal oscillator VCXO block is used as reference input to two synthesizers, FS4 and FS5, and a phase lock loop (PLL) block. The FS4 output mixes with the incoming signal to create an IF frequency. The IF chain output is one input to a PLL, whose reference input is the output of the VCXO. As in the interrogator, all frequency inputs are divided down so that the phase detectors in the frequency synthesizers compare phases of equal frequencies. The PLL adjusts the control voltage on the VCXO until lock is obtained. At this point, the output of FS5 will be frequency and phase locked to the incoming signal. The responder transmits an unmodulated signal to the initiator, which downloads the signal and compares its phase to that of the initiator's local crystal oscillator.

Timing of the system is illustrated in Figure 5.16. Transmission frames have a preamble defined according to a communication protocol that facilitates acquisition of desired signals. Data may be transmitted and received after the preamble, or a distance measuring routine may be conducted. During the distance-measuring DM period, an unmodulated carrier is transmitted. At t_1 the initiator transmits a packet, which is received by the responder. After allowing time for its FS4 and PLL loops to stabilize, at t_2 the responder opens the PLL output to the VCXO control line by a sample/hold switch [Figure 5.15(b)]. The VCXO control voltage is maintained essentially constant by a capacitor, and the VCXO output is open loop during the interval T_{hold}. In this state, phase coherency is maintained when the received signal ceases and the responder changes over to transmit mode at t_3. The interrogator's transmission should end shortly after the known instance, when the responder's hold order is issued and the PLL is open loop, earlier than t_3. Both sides need a given time for TX/RX changeover, after which the responder starts its transmission on the same hop channel as the interrogator, beginning with the preamble. It then must continue to transmit until the interrogator I and Q signals used for the distance-measuring algorithm are stable after lowpass filtering and are read at t_4. At the end of the responder's transmission, both sides move to the next hopping channel, with the interrogator transmitting and the responder receiving a new distance-measuring packet.

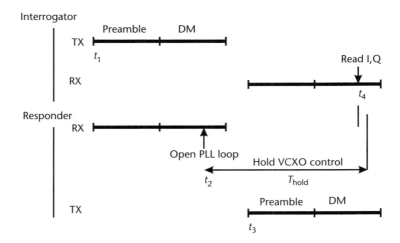

Figure 5.16 Multicarrier distance measurement system timing diagram.

The precision of the timing instances indicated on Figure 5.16 is not critical to the distance measurement process. It is important, however, that T_{hold} be maintained until after the interrogator reads the I, Q values for that particular frequency hop, that is, after t_4.

Figure 5.16 shows the period T_{hold} that the responder's reference frequency control loop is open. During this time, holding capacitor leakage and natural relative drift between the reference oscillators on both sides will cause an error in the phase measurement. The hold time should be as short as possible but its actual duration is a compromise between drift considerations and the need for narrow I/Q filter bandwidth and PLL loop bandwidth in order to reduce noise. Narrow bandwidths increase settling time.

The frequency hopping distance-measuring system just described can operate as a normal FHSS communication system, with specially defined DM packets included in the protocol. The hopping channel sequence is pseudorandom, and the measurements are put in order after a complete set of channels has been used. Thus, all benefits of FHSS for interference rejection are maintained for the DM function. ISM unlicensed bands can be used for distance measurement. For example, on the 2.4- to 2.483-GHz band, up to 79 or 80 hopping frequencies can be used, with 1-MHz spacing. These parameters are adequate for measurement of one-way distances up to 150m with resolution on the order of 1m in a multipath environment.

5.6.1   Phase Difference Measurements and Analogy to TDOA

Multicarrier phase analysis can be conveniently used to make distance difference estimates for finding target location. The method relates to the previously described multicarrier phase measurement procedures as time difference of arrival (TDOA) relates to time of arrival (TOA) (see Chapter 2, Section 2.2.4). The concept of locating a target by using differences in path lengths is based on the geometrical property that all positions of a target for which the difference of distances to a pair of fixed terminals is constant lie on a hyperbola (two dimensions) or hyperboloid (three dimensions). The intersection of the hyperbolas or hyperboloids created by different pairs of stations locate the position of the target. In TDOA, the fixed terminals have synchronized clocks. In a multicarrier distance measuring system, the local oscillators of the fixed terminals are coherent (have equal frequency and a constant phase relationship).

An implementation of a multilateral TDOA equivalent multicarrier phase system is shown in Figure 5.17. There is no responder, and the target transmits a constant CW or narrowband signal. Three geographically separated fixed stations receive the target transmission. Station 1 locks on to the received signal using a phase lock loop comprising a VCO and phase comparator with feedback control, indicated by a dashed line. The VCO output is distributed to the other two stations to serve as a phase reference. Stations 2 and 3 produce an estimate of the difference in path lengths, Δd_1 and Δd_2, from the target to each of them and station 1 in a manner similar to that of the interrogator receiver shown in Figure 5.15(a). Knowing Δd_1 and Δd_2, and the positions of the three fixed stations, the system can estimate the coordinates of the target. The calculations must effectively cancel out the phase shift of the reference source over the distance between Station 1 and Stations 2

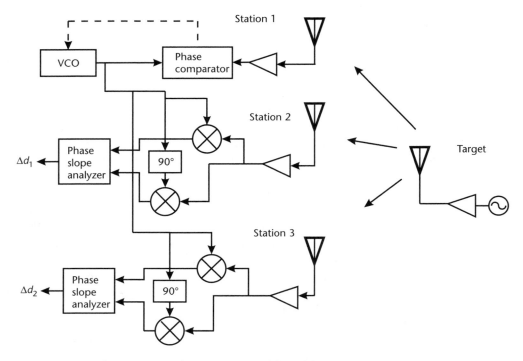

Figure 5.17 Multicarrier equivalent to TDOA multilateral location system.

and 3. Details of the TDOA location method are given in Chapter 7. Reference [5] describes a location system based on hyperbolic phase trilateration techniques and continuous wave phase measurements.

5.7 OFDM

We have seen how to estimate distance by measuring phase differences on hopping channel frequencies. Orthogonal frequency division multiplex (OFDM) is a modulation technique that spreads a data stream across many carrier frequencies transmitted simultaneously. It would seem therefore that multicarrier distance measurement could be carried out on an OFDM communication link. In OFDM conversion of data to a multicarrier signal in the transmitter and reconstruction of the data in the receiver are carried out using Fourier transforms on sampled signals. We have seen above that one way of extracting time delay information from multicarrier phase information is through the use of Fourier transforms. So multicarrier distance measuring techniques may be particularly applicable to OFDM.

Before getting into the details of distance measurement using OFDM, we first describe the principles of OFDM communication.

5.7.1 The Basics of OFDM

In OFDM, bits or small groups of bits individually modulate harmonically related subcarriers that are transmitted simultaneously. This process is illustrated in Figure

5.18. The phase of each subcarrier, which is constant during a symbol period, represents the binary value of 1 bit or the binary value of a subgroup of bits. The symbol period in the example of Figure 5.18 is four times the period of a data bit. By extending the symbol period, without reducing data rate, intersymbol interference in a multipath environment is reduced. The subcarriers are orthogonal so that there is no interaction or interference between them. Thus, the subcarriers obey the following expression for othogonality:

$$\int_0^T x_p(t) \cdot x_q(t) \cdot dt = 0 \quad p \neq q \tag{5.39}$$

where T is the symbol period. Two orthogonal tones are shown in Figure 5.19.

The lowest subcarrier frequency and the separation between the subcarriers, f_1, is a function of the sampling frequency f_s and the number of samples per symbol, N:

$$f_1 = \frac{f_s}{N} \tag{5.40}$$

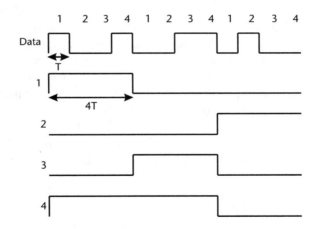

Figure 5.18 OFDM data bits allocated to separate subchannels.

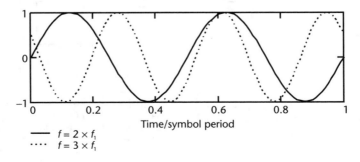

Figure 5.19 Orthogonal constant phase signals during one symbol period.

A complex vector represents the set of carriers in an OFDM signal. For each symbol a vector component has an amplitude and a phase angle that are determined by the bit or bits in the data stream that it represents. For example, if one bit of data is associated with one subcarrier, then that subcarrier component in the vector may have a magnitude of unity and phase of 0° or 180°, or equivalently a magnitude of ±1, depending on whether the data bit value is "0" or "1." If two bits are carried on a subcarrier per symbol, then QPSK modulation is used and the subcarrier phase may be 0°, 90°, 180°, or 270° according the binary value of the two bits. Not all of the N vector components are populated by a subcarrier and those that are not have a null value. An inverse fast Fourier transform (IFFT) of the vector of subcarriers creates a time domain vector representation of the symbol that is read out at the sample rate and upconverted to the transmission channel frequency band. The baseband time domain signal sample values $x(n)$ that result from the IFFT of the data vector X are expressed as:

$$x(n) = \frac{1}{N} \cdot \sum_{m=0}^{N-1} X(m) \left[\cos(2\pi m f_1 n t_s) + j \cdot \sin(2\pi m f_1 n t_s) \right] \quad (5.41)$$

where t_s is the sample time $1/f_s$.

Figure 5.20 is a block diagram of the OFDM transceiver. Examples of operational characteristics are taken from IEEE Standard 802.11a. In the transmitter (upper signal path), serial data, which includes error correction coding, is mapped to phase and amplitude values for each subcarrier. The frequency domain vector is transformed to time domain samples in the IFFT block. Following the IFFT, a cyclic prefix is inserted at the beginning of the transformed vector. This prefix, Figure 5.21, is a copy of the latter portion of the original time domain vector. For communication purposes, the cyclic prefix serves as a guard band to prevent multipath signal reflections of a previous OFDM symbol from overlapping the original part of the present symbol, included within the data symbol time T_D in Figure 5.21. The OFDM symbol with its cyclic prefix is upconverted onto an RF

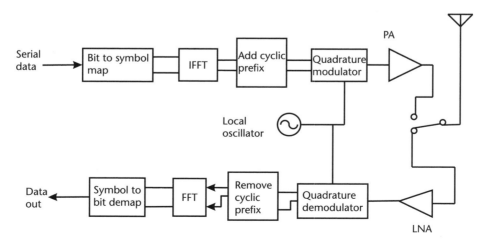

Figure 5.20   Basic block diagram of OFDM transceiver.

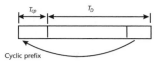

Figure 5.21 Creation of the cyclic prefix.

carrier frequency for radio transmission using quadrature modulation. Figure 5.22 displays simulated OFDM signals. The upper waveform shows BPSK signals in the frequency domain. Symbol data samples are arranged on both sides of the carrier frequency, which is indicated in the plot as a suppressed sample number 31. There are 52 active subcarriers among the 64 vector components of the signal vector in this example. Note that 11 of the 12 suppressed subcarriers are located at the high and low extremes of the frequency spectrum and create guard bands between channels. The twelfth suppressed subcarrier is the center frequency. The transmitted symbol samples in the time domain are shown in the bottom waveform. This is a complex wave represented by I and Q outputs, corresponding to the cosine and sine terms in (5.41). Only the magnitude of the signal is shown in the figure. The end 16 data bits are copied to the input and make up the cyclic prefix, so the complete symbol has 80 sample values.

On reception an operation that is the reverse of the procedure described for the transmitter is carried out (bottom data path in Figure 5.20). The incoming signal is downconverted coherently to quadrature components at baseband and

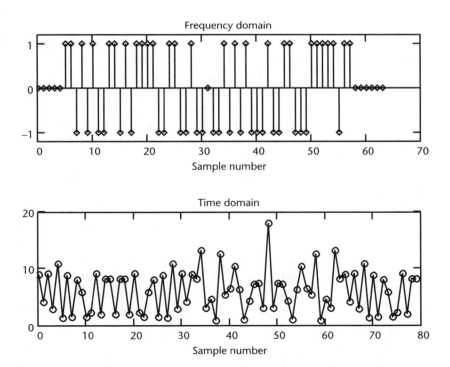

Figure 5.22 Frequency and time domain plots of OFDM signals.

sampled at rate f_s. The cyclic prefix is removed from the symbol data stream. A fast Fourier transform is taken on the remaining samples which results in a phase vector with number of elements equal to the original number of subcarriers. If the incoming signal is sampled over the exact same sampling window of the baseband signal that it was created from; that is, starting after the end of the cyclic prefix and at the beginning of the original baseband signal, the phase vector after the FFT will be exactly the same as the originally transmitted phase vector. However, if the sampling window in the receiver begins within the cyclic prefix, the phase in each element of the vector resulting from the FFT of the window samples will lag the phase of the corresponding element of the transmitted vector by an amount that is proportional to the time between the first sample of the received signal and the end of the cyclic prefix. This relationship between a time delay in the time domain and the phase in the frequency domain is known from a theorem in Fourier transform theory that may be stated as follows: If a signal $x(t)$ is delayed in time by t_0 seconds to produce a new signal $x(t - t_0)$ then the spectrum is modified by a linear phase lag of $j2\pi f t_0$, that is [6]:

$$x(t - t_0) \leftrightarrow X(f)e^{-j2\pi f t_0} \tag{5.42}$$

5.7.2 OFDM Distance Measurement

Distance measurement using OFDM is based on the linear relationship between the slope of the phase shift of symbol subcarriers and propagation time. The measurement process will be described with reference to block diagrams Figures 5.23 and 5.24, and the timing diagram of Figure 5.25. The system consists of an interrogator and a responder. A packet of OFDM symbols originating in the interrogator terminal is transmitted to the responder. It begins with a preamble to facilitate frequency and symbol timing (not shown in Figure 5.25). In the interrogator of Figure 5.23, a subcarrier phase vector is created as described in Section 5.7.1. Its components are forwarded in parallel to the IFFT block and to the resulting output is added the cyclic prefix. The sample clock times the output of a digital to analog converter that is upconverted in a modulator to the transmission frequency band. A symbol strobe, shown in Figure 5.23 as an input to the IFFT block, marks the beginning of each symbol.

In a distance measurement (DM) protocol, the DM symbols and their place in the packet are known to the responder. Referring now to the timing diagram, Figure 5.25, the interrogator aligns a periodic symbol strobe to the beginning of the first DM symbol, t_{ssi}, which continues at the symbol rate after the packet burst termination while the terminal changes from transmit to receive. A synchronized data strobe is produced at the end of the cyclic prefix, t_{ds}. The relationship between the clock and strobe pulses is shown in Figure 5.24. The number of sample clock pulses within a complete symbol period is indicated by M. The symbol period is the duration of the cyclic prefix, T_P, plus the time of the data symbol, T_D.

The responder downconverts the received signal and synchronizes its clock to the demodulated bits. It aligns an input symbol strobe to the beginning of the cyclic prefix at t_{ssr} (Figure 5.25) and proceeds to clock the signal into the buffer/store block of Figure 5.23, at the synchronized sample clock rate of the interrogator.

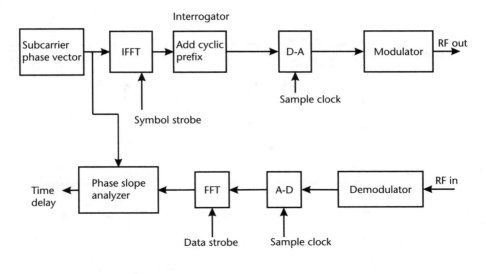

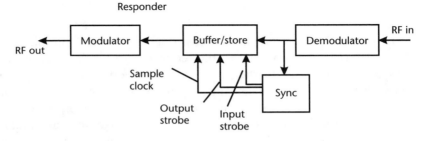

Figure 5.23 Interrogator and responder in OFDM distance measuring system.

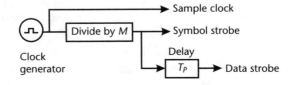

Figure 5.24 Strobe and clock generation in OFDM distance measuring system.

When the entire DM symbol has been sampled and the samples stored in a memory register, the responder changes over from receive mode to transmit mode. After a period of time that is equal to an integer number, n, of symbol length times after the symbol strobe, that is, at $t_{ssi} + nT_s$ in Figure 5.25, the responder clocks out the samples from the buffer/store unit and transmits them to the interrogator. The value of n takes into account the time required for the interrogator to complete transmitting the OFDM burst and change over to receive mode. In a distance measurement protocol, it should not be necessary to actually sample the incoming bits, just to mark the instant of the start of the cyclic prefix. The responder can know in advance the symbol that the interrogator will send it and to maintain a stored copy of the symbol sample sequence. Therefore, after detecting the beginning

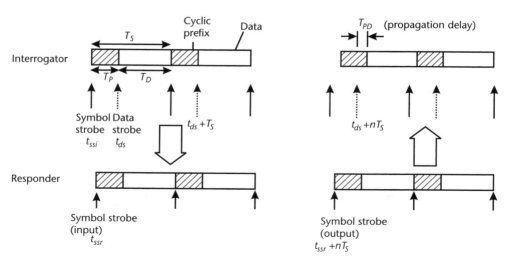

Figure 5.25 OFDM distance measurement timing diagram.

of the received symbol sequence, it does not have to sample that sequence and it transmits the stored copy of the sequence after a delay of an integral number of symbol periods as in the previous description. An advantage of using a stored symbol is that the retransmitted symbol will not be contaminated by noise or interference. However, it can be used only when the distance-measuring symbol is constant and is established in advance. A disadvantage in using the stored symbol is that the exact position of t_{ssr} must be determined, which is not the case when the incoming signal is buffered for later readout.

On the interrogator side, after sending the OFDM burst, the terminal changes over from transmit mode to receive mode while maintaining the data strobe clock uninterrupted. It receives the return OFDM burst retransmitted by the responder and clocks demodulated signal samples through an analog to digital converter, (A-D in Figure 5.23) to the FFT block. The FFT operation begins at a data strobe, $T_{ds} + nT_s$ (Figure 5.25), n symbol times plus the period of the cyclic prefix after the original symbol strobe of the interrogator at t_{ssi}. Symbol sampling is carried out for a duration of T_D. Due to the two-way propagation delay of the signal, the sample window for the FFT in the interrogator receiver commences before the start of the data portion of the symbol, that is, during the cyclic prefix. The value of this delay, T_{PD}, which equals the time from the beginning of the first sample to the end of the cyclic prefix, can be determined from the phase difference between the subcarriers that were transmitted from the interrogator to reception of the signal retransmitted from the responder. From the argument of each element, or subcarrier phase, of the output of the FFT is subtracted the argument of the corresponding element of the frequency domain data vector of the originally trans-mitted signal. This operation takes place in the phase slope analyzer block in Figure 5.23. The phase versus angular frequency slope of the resulting difference vector is the propagation delay, from which the distance between interrogator and responder can be calculated. Instead of measuring the phase slope directly, the time delay is preferably found by taking the IFFT of the phase difference vector, which is expressed as complex elements. The ideal correspondence between the phase difference vector and the time delay resulting from the IFFT is shown in the Fourier transform shifting theorem, (5.42).

A plot of the IFFT of the phase difference vector can be examined and analyzed to separate the direct signal from multipath. The phase difference IFFT output of the interrogator receiver in an OFDM distance measurement system simulation is shown in Figure 5.26. A direct path and three echoes were used in the simulation. The parameters of the returns are shown in Table 5.2. The direct path delay over Path 1 is τ. The same parameters were applied to both the forward and reflected transmission.

The following are the relevant parameters of the OFDM system:

- Data samples per symbol: 64;
- Active subcarriers: 52;
- Cyclic prefix: 16 samples;
- Sample rate: 20 Mbps.

In addition, the return signal at the interrogator receiver is oversampled by a factor of 4; that is, 256 samples per data period T_D.

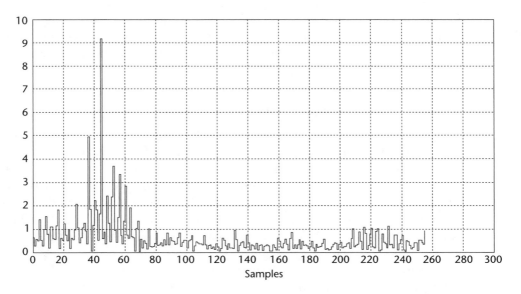

Figure 5.26 OFDM distance measurement simulation result.

Table 5.2 Multipath Parameters for OFDM Distance Measuring Simulation Example

	Delay	Rel. Strength	Phase Rotation
Path 1	τ	0 dB	0°
Path 2	1.2 τ	−10 dB	−10°
Path 3	1.3 τ	−6 dB	−45°
Path 4	1.5 τ	−7 dB	20°

It is necessary to determine what criteria to use to decide which peak is the true direct path. In Figure 5.26, the direct path is the largest peak. The two-way propagation delay between the interrogator and the responder is the sample number of the peak (N_{sample}) times the period of the sample clock ($1/f_s$), divided by the oversampling factor. Thus, in this example, the true one-way range is given by the following expression:

$$range = \frac{c}{2} \cdot N_{sample} \cdot \frac{1}{f_s} \cdot \frac{1}{oversampling\ factor} \tag{5.43}$$

where c is the speed of light. In Figure 5.26, the direct peak occurs at $N_{sample} = 44$, and using (5.43) range = 82.5m.

Normally it cannot be presumed that the direct path is dominant and an algorithm that manipulates the phase difference IFFT output has to be developed to determine the propagation time of the earliest received multipath return in order to provide an estimate of the true range.

There are several basic differences between distance measurement on frequency hopping channels and OFDM. In OFDM, the distance measurement data is taken from modulated waveforms whereas in the case of frequency hopping the carriers are unmodulated during the measurement time. In the frequency hopping method, accuracy may be increased by reducing the I/Q lowpass filter bandwidth—decreasing signal bandwidth. The most straightforward means for increasing OFDM DM accuracy is to increase the sampling rate in the interrogator receiver. When the sampling rate is increased without increasing the number of subcarriers the bandwidth remains the same but resolution is improved.

5.7.3 Location Based on OFDM Distance Measurement

The principle of distance measurement based on OFDM can be applied to determining the position coordinates, that is, the location of an OFDM transmitter. Two methods of OFDM location determination are described here. The first method described below is particularly useful because the OFDM transmitter whose location is to be determined may be a standard OFDM communicating device that has no special facilities for the distance measurement function. The only requirement is that the symbol length T_S must be equal to an integral number times the length of the cyclic prefix T_P.

Figure 5.27 shows an example of terminal deployment for finding location of a target. TX is the OFDM transmitter whose location is to be found. It may be regarded as a client terminal, mobile or fixed, in a communication network. RX1, RX2, and RX3 are fixed access points in the network each of whose coordinates are known. All access points have a common data strobe pulse train whose rate is the inverse of the cyclic prefix length T_{CP}. Strobe pulses occur simultaneously at each access point. Any differences in the time of occurrence of the pulses at the access points can be canceled out by knowledge of strobe pulse distribution time delays.

Timing diagram Figure 5.28 shows reception at an access point receiver, RX1, RX2, or RX3, of a symbol transmitted by TX. The time of arrival of each symbol

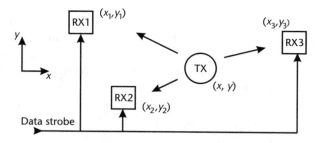

Figure 5.27   OFDM location system layout in two dimensions.

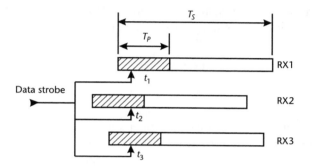

Figure 5.28   OFDM location timing diagram.

is a direct function of the distance between TX and an access point receiver. t_1, t_2, and t_3 in the diagram are the time delays to the end of the cyclic prefix relative to the data strobe. Each receiver composes the FFT block vector output in the receive channel of Figure 5.23 using, at the input to the FFT block, the data sampled at the instant of the data strobe occurring during the cyclic prefix. The three access points in this example are coordinated so that all measurements start at the same data strobe occurrence and on the same transmitted symbol. One of the three access point receivers, say, RX1, makes its vector of FFT received data available to the two other receivers, RX2 and RX3, for use in the phase slope analyzer block instead of the interrogator transmitter subcarrier phase vector shown in Figure 5.23. The calculated time delays at the output of the phase slope analyzers of RX2 and RX3 are therefore the time differences of arrival of the target transmitter transmission over the paths between each of the receivers RX2 and RX3, and the designated reference receiver RX1.

   In a two-dimensional situation, as depicted in Figure 5.27, the location of TX is expressed as coordinates (x, y). Known locations of RX1, RX2, and RX3 are (x_1, y_1), (x_2, y_2), and (x_3, y_3). The unknown coordinates x and y can be found by solving the nonlinear equations that express the path length differences which are given by the time differences of arrival times the speed of light, c. These equations are:

$$\sqrt{(x - x_2)^2 + (y - y_2)^2} - \sqrt{(x - x_1)^2 + (y - y_1)^2} = c \cdot t_{1,2} \qquad (5.44)$$

$$\sqrt{(x - x_3)^2 + (y - y_3)^2} - \sqrt{(x - x_1)^2 + (y - y_1)^2} = c \cdot t_{1,3}$$

where $t_{1,2}$ and $t_{1,3}$ are the time delay difference measurements at access point receivers RX2, and RX3. The location of TX is found by solving these equations for x and y. Similarly, three-dimensional locations, having coordinates (x, y, z) can be determined from the measurements from at least four access points by extending the above equations by an additional equation (or equations for additional redundant access points) and adding under the square root operations the term $(z - z_i)^2$ where subscript i refers to the particular access point. Chapter 7 gives details on finding location by the TDOA method.

A second OFDM location determining method is derived from the case where absolute distance measurements are made using the principles explained in Section 5.7.2. In the two-dimension situation only two access points are necessary and the unknown coordinates (x, y) are solved from equations:

$$(x - x_1)^2 + (y - y_1)^2 = (c \cdot t_1)^2 \qquad (5.45)$$

$$(x - x_2)^2 + (y - y_2)^2 = (c \cdot t_2)^2$$

where t_1 and t_2 are the measured propagation time delays at the access points. For three-dimension location, at least three access points are required and coordinates (x, y, z) may be found by solving three equations. Synchronized strobes are not required. As an example of use, this method can be applied to a stock management system where dedicated OFDM measurement responder tags are attached to articles that need to be tracked, such as hospital equipment or merchandize in a warehouse.

5.7.4   Resolution of OFDM Distance Measurement

The smallest interval of two-way propagation time delay that can be detected at the interrogator is the sample period, or the inverse of the sampling frequency. For example, the sampling frequency of OFDM wireless LAN according to IEEE specification 802.11a is 20 MHz and the sample period is 50 ns. This corresponds to a one-way distance resolution between interrogator and responder of 7.5m. One way to increase distance resolution is to raise the frequency of the sample clock at the receiver A-D block in the interrogator receiver, Figure 5.23, by an integer multiplier N. This will necessitate upsampling the reference phase vector by N before entry to the phase slope analyzer and taking a corresponding larger FFT. The sample length of the FFT will be the length of the original subcarrier phase vector times N.

A second way to obtain higher distance resolution and accuracy by increasing the measurement time involves using a delay line with multiple taps. This method, which does not use a higher sampling rate, is carried out by taking a number of measurements where for each measurement the sampling clock and symbol strobe are delayed by a fraction of the sample period. The timing method is shown in Figure 5.29. The sample clock and data strobe are applied to the receiver A-D and FFT blocks of Figure 5.23. Each subsequent measurement is taken with the clock generator pulse of Figure 5.29 delayed by an additional fraction of the time period by switching the delay line tap. For example, in a system having a sample rate of 20 MHz, the sample period is 50 ns. To increase resolution by 10, 10 measurements

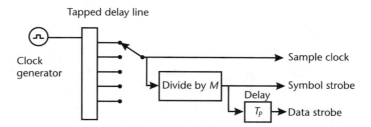

Figure 5.29   Tapped delay line for improving distance resolution.

are taken and on each subsequent measurement the sample pulses are delayed by an additional 5 ns. The final distance measurement result is calculated from the average of the 10 measurements.

The maximum nonambiguous distance that can be measured corresponds to the length of time of the cyclic prefix, T_P. For example, in OFDM of IEEE 802.11a the cyclic prefix length is 800 ns, corresponding to a distance between interrogator and responder of 240m. The maximum prefix length that can be used is equal to the data length, T_D. In the IEEE 802.11a protocol $T_D = 3.2$ μs, corresponding to a maximum distance of 960m.

5.8   Conclusions

The multifrequency method is appropriate for use with communication systems implementing narrowband hopping channels under FHSS, as an alternative to time of arrival (TOA) ranging. It demands different trade-offs compared to epoch time measuring systems. Whereas time-of-arrival precision is increased by increasing bit rate and therefore bandwidth, the phase measurement time of flight precision is improved by narrowband filtering. Reduced bandwidth, however, translates to increased measurement time, so in comparing the two methods, the high bandwidth techniques of DSSS and ultrawideband used in time domain distance measurement result in reduced ranging time compared to multifrequency phase comparison distance measurement. On the other hand, the high bandwidth of UWB and DSSS means reduced SNR, unless multiple measurements are averaged, increasing the ranging time. DSSS data demodulation or ranging is inherently an averaging mechanism, as timing is established by correlating over a symbol period or its multiple. Thus for a given signal-to-noise ratio, or operational communications range, performance of both methods are equivalent in terms of measurement time, and the choice between them is a question of implementation convenience for a given application. Measures for combating multipath interference differ for the time and phase methods, but theoretically achievable results should be comparable.

Distance measurement based on OFDM signals is in a way a hybrid method, having aspects of accurate epoch timing and phase comparison. OFDM requires coherent demodulation and sample synchronization, while the range determination itself is based on phase comparison and not reading a timer. Both FHSS and OFDM techniques are susceptible to multipath propagation, although when a strong direct path is present a good propagation time estimate can be determined using Fourier transform techniques.

Similar location techniques to those used with TOA methods are applicable to multifrequency phase difference methods as well. Position estimation by triangulation, for example, uses the two-way ranges determined by the multifrequency method between fixed terminals and a target. We have shown in Section 5.7.3 a convenient use of OFDM for TDOA positioning.

Multifrequency ranging techniques have not been employed to the extent of the other methods, based on time of arrival and received signal strength. However, they may be the most appropriate for adding a ranging dimension to existing FHSS and OFDM communication systems, as well as for accurate ranging only applications where a high rate time base is not available.

References

[1]   Palmer, R. J., "Test Results of a Precise, Short Range, RF Navigational/Positional System," *IEEE Vehicle Navigation and Information Systems Conference*, Ontario, Canada, September 11–13, 1989.

[2]   Proakis, J. G., *Digital Communications*, 3rd ed., New York: McGraw-Hill, 1995.

[3]   Davenport, Jr., W. B., and W. L. Root, *An Introduction to the Theory of Random Signals and Noise*, New York: McGraw-Hill, 1958.

[4]   MATHCAD, http://www.ptc.com.

[5]   Feuerstein, M. J., T. Pratt, and Y. J. Beliveau, "A Precision Automatic Vehicle Location System for Use in Construction Automation," *IEEE Vehicle Navigation and Information Systems Conference*, Ontario, Canada, September 11–13, 1989.

[6]   Carlson, A. B., *Communication Systems*, New York: McGraw-Hill, 1968.

CHAPTER 6

Received Signal Strength

Chapters 3, 4, and 5 described distance measurement and location methods that are based on propagation time—the time for electromagnetic radiation to propagate from a transmitter to a receiver. In this chapter, methods of relating distance to received signal strength (RSS) are discussed. On the average, signal strength at a receiver decreases as distance from the transmitter increases. If the relationship between signal strength and distance is known, analytically or empirically, the distance between two terminals can be determined. When several base stations and a target are involved, triangularization can be applied to determine the target's location.

6.1 Advantages and Problems in RSS Location

RSS has several advantages over the TOF methods. It can be implemented on an existing wireless communications system with little or no hardware changes. All that's needed is the ability to read a RSSI (received signal strength indicator) output that is provided on virtually all receivers, and to interpret the reading using dedicated location estimation software. The modulation method, data rate, and system timing precision are not relevant. Coordination or synchronization between the initiator and the responder for distance measurement are not required. Thus location capability can be added to a wireless system for very low incremental cost.

On the other hand, there are specific problems in implementing location awareness with the RSS method. Because of large variations of signal strength due to interference and multipath on the radio channel, location accuracy is generally less than what can be achieved using TOF methods. Propagation is location/environment specific, and system software usually has to be tailored to the place where the system is being used. Often, as will be shown later, a specific database must be created for a given location. In order to achieve a useful accuracy in a location system, many more fixed, or reference terminals, are required than the minimum number needed for triangulation. Orientation of a target as well as its location related to nearby objects will have an effect on the location estimation.

There are two basic classes of systems that use RSS to estimate location: those that are based on known radio propagation analytic relationships, and those that involve searching a database that is composed of measured signal strengths in a location specific survey. The latter class is often referred to as fingerprinting. A third class can be defined as a combination of the first two—a database is formed from the use of analytic equations or derived from ray tracing software.

6.2 Propagation Laws

The way signal strength changes as a function of distance from a radiating source is a function of the environment. The simplest and most exact formulation of that function is applicable to free space. Any other environment contains objects that reflect, absorb or obstruct, or scatter the electromagnetic wave, forcing a modification of the free-wave signal strength versus distance relationship and the introduction of a probabilistic term to account for the fact that the environment cannot be described exactly or changes with time.

6.2.1 Free Space

In free space, the parameters that directly affect the relationship between received power P_r and distance d at wavelength λ are included in the Friis equation:

$$P_r = \frac{P_t G_t G_r \lambda^2}{(4\pi)^2 d^2} \tag{6.1}$$

G_t and G_r are transmitter and receiver antenna gains. Note that the receiver cannot calculate the distance to the transmitter only from the received power, but it must be informed of the transmitter's radiated power—$P_t G_t$—either from previous knowledge or a message from the transmitter. This is analogous to the situation in a TOF system where a receiver must know an epoch time of transmission in order to find the one-way propagation time. Consequently, distance to a rogue transmitter, for example, cannot be determined by a single receiver without some cooperation from the target terminal.

Equation (6.1) can be made more convenient for purposes of comparison as well as simplified by expressing it as the inverse of the numerical path loss, or path gain, PG. Path loss is the attenuation of the signal as it propagates between transmitter and receiver. We use path gain instead of path loss in order to show more directly the effect on received signal strength. Numerical path gain is the ratio of the received power to the transmitter radiated power,

$$PG = \frac{P_r}{P_t G_t G_r} = \left(\frac{\lambda}{4\pi d}\right)^2 \tag{6.2}$$

6.2.2 Free-Space dB

It is usually more convenient to work with logarithmic expressions, for which path gain in decibels is

$$PG_{dB} = 20\log\left(\frac{\lambda}{4\pi d}\right) \tag{6.3}$$

In free space, when transmitted power and antenna gains are known, distance can be determined with high accuracy from the received signal strength using (6.3).

However, in all other communication links, objects, including the ground, in the vicinity of the transmission path change the relationship between received power and distance. The received power will be a vector sum of signals from the transmitter arriving over different path lengths because of reflections from nearby objects and partial blocking by materials in the signal path. The resulting received power may be greater or less than the line-of-sight signal over the transmission path. When the reflecting objects are moving in respect to the link terminals, the received power will change with time. In addition large obstacles such as buildings, walls, or floors that are present on the line-of-sight path attenuate the direct signal and reduce the received power.

6.2.3   Open Field

A plot showing how received signal strength varies with distance in the presence of one reflector, the ground, is shown in Figure 6.1, which also shows free space path gain for comparison. Frequency is 2.4 GHz and both the transmitting and receiving antennas are vertically polarized and 1.5m high. Vertical antennas are most commonly employed on 2.4-GHz short-range devices because they are nondirectional and most convenient to attach to small products. Within the distance span shown in the plot, 100m, the path gain, and consequently the received signal strength, varies significantly from the free space value as expressed in (6.3). As a mobile terminal recedes from a fixed terminal, the signal experiences variable fading, and within short distances the received signal strength grows while the range increases. The mean value however follows closely the free-space curve. Over larger distances than are shown in Figure 6.1, the variations over small distance increments decrease and the open field signal strength is consistently below the that of free space.

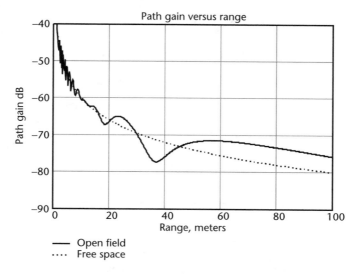

Figure 6.1   Open field and free-space propagation path gain at 2.4 GHz. Polarization is vertical and transmitting and receiver antenna heights are 1.5m.

When the range (d) axis is a logarithmic scale, the mean signal strength curve can be approximated by two linear segments that meet at some distance d_0. The path gain curve shown in Figure 6.1 is plotted in Figure 6.2 with a logarithmic range axis and maximum range extended out to 1,000m. To the left of the vertical dashed line marked as d_0, the log-log plot has an average slope of −2, representing a distance exponent of 2, as in free space, and the segment to the right of d_0 has a slope of −4, showing dependence of a distance exponent of 4. The plot can be expressed approximately by

$$PG_{dB} = -20 \, \log\left(\frac{4\pi d_0}{\lambda}\right) - 20 \, \log\left(\frac{d}{d_0}\right) \qquad\qquad d \leq d_0 \qquad\qquad (6.4)$$

$$PG_{dB} = -20 \, \log\left(\frac{4\pi d_0}{\lambda}\right) - n \cdot 10 \cdot \log\left(\frac{d_0}{d}\right) \qquad\qquad d > d_0$$

where n is the exponent of the inverse of the distance when $d > d_0$. The path gain parameters are wavelength λ, d_0, and n. In the case of open field propagation, $n = 4$ and d_0 can be approximated by

$$d_0 = (12h_1h_2)/\lambda \qquad\qquad (6.5)$$

where h_1 and h_2 are the heights of the terminal antennas. In Figure 6.2, h_1 and h_2 each equal 1.5m and λ equals 0.125m, resulting in a value for d_0 of 216m.

6.2.4 Logarithmic Approximation

Curve approximations expressed by (6.4) with plots similar to Figure 6.2 can be made when there are other reflections in addition to ground. d_0 and n can be

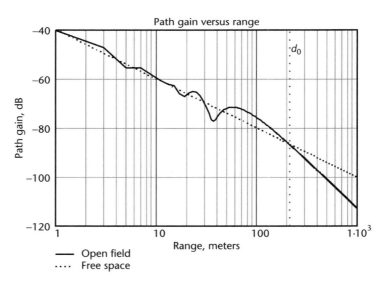

Figure 6.2 Open field and free space propagation path gain at 2.4 GHz with a logarithmic scale on the range axis. Polarization is vertical and transmitter and receiver antenna heights are 1.5m.

estimated empirically by survey measurements. The slopes and intercepts of (6.4) are calculated by least square regression from the empirical data, choosing d_0 by estimation from observing the data.

A simplified propagation model for indoor environments over a range of 0.5m up to several hundred meters is shown in (6.6) and plotted in Figure 6.3. The model is for a frequency of 2.45 GHz and has been suggested for use in wireless personal area networks [1]. The path gain estimation is for free space propagation from 0.5m up to 8m. Beyond 8m, the estimated path gain has a slope of −3.3 ($n = 3.3$).

$$PG_{dB} = -40.2 - 20 \, \log\left(\frac{d}{1m}\right) \qquad 0.5m \leq d \leq 8m \qquad (6.6)$$

$$PG_{dB} = -58.5 - 33 \, \log\left(\frac{d}{8m}\right) \qquad d > 8m$$

6.2.5   Randomizing Term X

A number of models in addition to (6.6) have been suggested for indoor propagation [2]. Due to the wide variation of propagation conditions in indoor environments, no one formulation can adequately predict received signal strength in every installation. The following factors affect propagation and cause deviations from the various propagation relationships that have been suggested:

- Multipath propagation that depends on the position of the transmitter and receiver relative to floor and ceiling, partitions and furnishings;

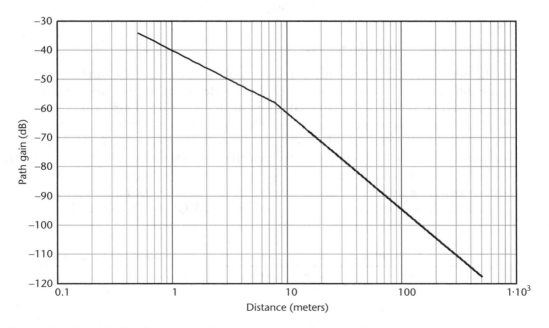

Figure 6.3   Example of path gain curve for indoor propagation at 2.45 GHz.

- Shadowing effect of building materials and other objects in the propagation paths;
- Antenna heights and relative polarization;
- Transmission frequency;
- Moving objects, specifically people, in the vicinity of the transmission paths.

In order to keep the propagation formula simple and also indicate deviations from what may be considered a mean large scale value, a term indicating randomness due to any of the factors listed above is added to (6.4) and shown in (6.7):

$$PG_{dB} = 20 \, \log\left(\frac{\lambda}{4\pi d_0}\right) + 10 \cdot n \cdot \log\left(\frac{d_0}{d}\right) + X_\sigma \qquad d > d_0 \qquad (6.7)$$

X_σ is a random value in decibels having a standard deviation of σ. Examples of the variation of n and σ with environment and frequency are shown in Table 6.1 [3].

As mentioned, environmental conditions change with time, and different transmission paths, even in a similar locality, have different parameters. Thus, the received power is a random variable and in order to attain desired distance or location accuracy, averaging methods are used, based on multiple measurements.

The parameters d_0, n, and σ can be found for a particular environment and frequency by taking a set of measurements of signal strengths at known ranges at various positions and times and then using the data to make a least squares estimate of those parameters to fit the curve of (6.7). First, measurement data for range greater than a likely value of d_0 should be used to find n and σ, then the short-range data can be used to find a likely value of d_0.

6.2.6 Outdoor Area Networks

The details of range predications for outdoor mobile and fixed wireless networks are different than those of the indoor systems described above, but they generally are based on log-linear approximations in the form of (6.7). Empirical models have been proposed that apply to specific frequency bands and whose parameters are applied in a manner that depends on terrain or the degree of building density, described as large city, medium city, suburban, or open areas [4]. As an example, one of the models, designated as Stanford University Interim (SUI) Model, is described here briefly. It was developed for the IEEE working group 802.16 for fixed wireless access systems in the band from 2.5 GHz to 2.7 GHz. The systems

Table 6.1 Variation of Propagation Parameters with Environment and Frequency

Environment	Frequency (MHz)	Exponent n	Variance σ (dB)
Retail store	914	2.2	8.7
Office, hard partition	1,500	3.0	7.0
Office, soft partition	900	2.4	9.6
Factory, line of sight	1,900	2.6	14.1

are deployed with base station terminals (BS) and customer premises equipment (CPE). The model refers to three types of terrain and is applicable to suburban environments. Terrain type A has highest path loss and is characterized as hilly terrain with moderate to heavy foliage. Type C has minimum path loss and is based on flat terrain with light tree density. Type B is for an intermediate terrain with path loss between that of A and C. The pass loss equation for the model is:

$$PL = 20 \log\left(\frac{4\pi d_0}{\lambda}\right) + 10n\log\frac{d}{d_0} + X_f + X_h + s \qquad \text{for } d > d_0 \qquad (6.8)$$

d is the distance between two terminals and $d_0 = 100$m. s is a lognormal distributed factor that accounts for shadowing due to trees and other objects, with a value between 8.2 and 10.6 dB. n, the path loss exponent, is calculated from the expression

$$n = a - bh_b + c/h_b \qquad (6.9)$$

where h_b is the base station antenna height above ground, between 10 and 80m, and constants a, b, and c depend on the terrain type as shown in Table 6.2.
   X_f is a correction factor for the operating frequency f in megahertz:

$$X_f = 6.0 \log\left(\frac{f}{2,000}\right) \qquad (6.10)$$

The CPE antenna height h_r correction is applied to the model as X_h, which depends on the terrain type:

$$X_h = -10.8 \log\left(\frac{h_r}{2,000}\right) \qquad \text{for terrain types } A \text{ and } B \qquad (6.11)$$

$$X_h = -20.0 \log\left(\frac{h_r}{2,000}\right) \qquad \text{for terrain type } C$$

As in the case of indoor propagation models, the outdoor models generally produce a range of path loss estimations in a given environment. Therefore, they should be applied only under the conditions for which they were developed. When possible, empirical measurement sampling should be done to confirm the applicability of a given model in a particular situation and to assess the range or location accuracy that can be expected from it.

Table 6.2 Numerical Values for the SUI Model Parameters

Parameter	Terrain A	Terrain B	Terrain C
a	4.6	4.0	3.6
b (m^{-1})	0.0075	0.0065	0.005
c (m)	12.6	17.1	20
Source: [4].			

6.2.7 Path Loss and Received Signal Strength

As discussed previously, range is associated with path loss or path gain. In order to relate the measured received signal strength to distance, through the path loss or path gain expressions, radiated power and receiver antenna gain must be known. Transmitter power into the antenna, and transmitter and receiver antenna gains, are included in (6.1). In logarithmic terms, using decibels, path gain, PG_{dB} as a function of received signal strength is:

$$PG_{dB} = P_r - (P_t + G_t + G_r) \qquad (6.12)$$

where P_r is received signal strength, P_t is transmitter power to the antenna, and G_t and G_r are transmitter and receiver antenna gains, all in decibels. Path gain is the negative of path loss in decibels.

6.3 RSS Location Methods

Generally the technique of RSS is used for position location, not for one-dimensional distance estimation. Examples are handset location in a cellular network, allocating peripherals in WLAN, and location awareness in large-scale sensor networks. These systems deploy a number of fixed terminals with known coordinates in the detection area. Multilateral or unilateral modes may be used. In a multilateral system the mobile target transmits beacons that are received by each of the fixed base station terminals. A network-administrated unit can then estimate the target location from the received signal strength at each of the base stations. In the unilateral case the target computes its own location from received signal strengths from each of the base stations, and the knowledge of base station locations and radiated powers. Among the advantages of the multilateral arrangement are the availability of larger computational power and database capacity in the fixed infrastructure and the fact that the target needs no knowledge of specific location system parameters. A disadvantage is that the system administrator tracks the target's location, which may be considered as a violation of the target's privacy. The unilateral method may be preferred when location knowledge must be accessible only to the target itself. Also, it may accommodate a limitless number of targets simultaneously since computations are distributed among the multiple targets. The target may have to download specific base station parameters and an area database for each different area that needs location services.

There are basically two classes of techniques that are used for finding the location of targets using received signal strength. One class employs triangulation to find location from estimated distances between the mobile terminal and a number of fixed stations with known positions. The second class of location techniques using RSS involves matching real time signal strength measurements with database entries accumulated during a previous site survey.

6.3.1 RSS Location from Range Estimations

Distances are found from signal strength using formulas for propagation as discussed above. Location coordinates can be calculated based on range estimations acquired using a propagation model such as (6.6). The method of finding location from the geometry of the intersection of circles (two dimensions) or spheres (three dimensions) is the same as that used in TOF ranging location systems. Environmental conditions may be accounted for by choosing the propagation law parameters that are most appropriate to the area where the system is used, but no assistance is made from a previously prepared database. For most environments, location errors are significantly greater for RSS ranging and geometric location than with TOF methods.

An example of the use of one-dimensional distance measurement for location estimation is provided by smart sensor distributed networks. Such networks have ad hoc or peer-to-peer communication links that do not relate directly to fixed based stations with known coordinates and therefore the unilateral or multilateral definitions are not applicable. Location awareness for each sensor may be limited to knowledge of position relative to neighboring sensors. In this case the position information is relative, since no absolute location coordinates are available. Each sensor must have a capability of one-dimensional distance measurement. Ranges between three sensors constitute the known lengths of sides of a triangle whose angles can be calculated using the relationships between sides and angles of plane triangles. When some of the sensors do have absolute coordinates, it is possible for a system host to estimate the absolute location of all sensors after obtaining the relative position data that they have acquired in relation to nearest neighbors. One-dimensional RSS methods are attractive for sensor systems since hardware costs are low compared to other methods and signal bandwidth is not an important factor in performance.

6.3.2   RSS Location Based on Database Comparison

As compared to location determination based on RSS range measurements, better results may be achieved by comparing a set of signal strengths between target and base stations acquired in real time with signal strength measurements taken previously off-line at known locations throughout the coverage area. The techniques that employ database comparison are called fingerprinting, pattern recognition, or pattern matching. The database is applicable only for the particular site where it was created, and physical changes that affect radio propagation at the site may require creating a new database. The database comparison location technique is used for indoor and outdoor applications.

To illustrate the database search method of target location, we use a WLAN network as an example. The network is established in an office area with a floor layout shown in Figure 6.4. The area has corridors and partitions making up work cubicles. The squares mark locations where survey measurements were taken and the asterisks are access points (APs). Coordinates are referenced to the lower left corner, as marked.

The object of the location system is to estimate the coordinates, (x, y), of a terminal that enters the network. Although unilateral or multilateral computations may be used, we will assume that the target is the transmitter and the network infrastructure is responsible for calculating the location coordinates (multilateral). There are several uses that can be made of the system. It can be used to confirm

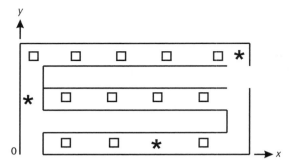

Figure 6.4   Office layout of location network.

identification of the client and to prevent the possibility of intrusion into the system of someone outside the protected area. Also, print requests can be directed to a printer that is closest to the target. It can also track a mobile target around the work area.

The location estimation process has two measurement phases. The first phase, called the off-line or survey phase, is the creation of a database. During the survey, signal strength is measured at each of the APs when a mobile WLAN terminal is transmitting from the survey reference points, indicated by squares on the floor layout of Figure 6.4. At each point the mobile unit, a laptop computer, transmits while it is oriented in four different positions: facing right and left of the x-axis and up and down on the y-axis. This is necessary because the radiation from the laptop is not truly omnidirectional and is dependent upon the antenna location on the computer and whether the operator's body is blocking line of sight between it and each AP. Signal strength varies over time due to small movements of the mobile terminal and movements of objects in the propagation paths, which may be people in the office in this example or the motion of trees and vehicles in an outdoor scenario. Therefore, the raw data for each reference point contains repeated signal strength measurements from a series of transmissions to all access points in range. The nature of the database that is created from the raw data depends on the comparison method, to be described in the following sections. Information components in the database are identified with the reference position to which the raw data is associated.

Instead of basing the raw data on actual measurements carried out at each reference position, the data can consist wholly or partially on propagation laws whose parameters are estimated by fewer sample measurements over the coverage area. Ray tracing software can also be employed when detailed construction or topographical information is available.

The second phase of the location procedure is the real-time online signal strength measurement process initiated by the location client when the mobile target's location is to be estimated. Signal strengths of target transmissions are recorded by all access points in range. The set of signal strengths acquired in this stage are compared with the database components associated with all reference points, and the result specifying the reference point or points of best match is used to indicate the estimate of target location.

The advantage of the database estimation method is that it is based on actual path loss at points near the target location and therefore unknown factors of shadowing and multipath are bypassed and affect only minimally the location estimation. However, unknown transmitter power, receiver signal strength indicator calibration, antenna orientation, and target blocking contribute to the uncertainty of the location position result. The biggest disadvantage of the method is the requirement for a site-dependent database that may be time consuming and expensive to create and cannot be reused in a different environment.

Several different ways for comparing real-time data measurements with the database have been developed. We describe two of them: that based on minimum Euclidian distance, referred to as the nearest neighbor method, and a statistical method using Bayesian inference.

6.3.2.1   Database Comparison by Search Nearest Neighbor

First we concentrate on the makeup of the database that is created in a survey phase of the location system installation. In our example, five signal strength measurements are taken at each AP for each of the four orientations at each survey point, for a total of 20 measurements for each point. The five signal strengths for each orientation are averaged to produce the mean. Thus, each element of data in the base is represented by the following vector

$$V_n = (x, y, p, s_1, s_2, \ldots, s_k, \ldots s_K)^T \qquad (6.13)$$

whose components are the location coordinates x, y, the orientation p, and mean signal strengths s_k. k represents the AP where the measurement was taken, K is the number of access points—five in this example—and n is the index of the survey reference point. Superscript T, for transpose, makes the expression equivalent to a column vector. If there are N survey points in the measurement area, including the four orientations for each coordinate, the database will contain N vectors. The number of individual reference locations equals $N/4$.

During the real-time measurement phase, when a station signs onto the network each AP records its signal strength and time the measurement is taken. A location coordinator (LC) that is part of the network infrastructure receives the measurements from the APs and associates those that occurred at the same time. Then the location coordinator checks the previously prepared database vectors for an entry of signal strengths that most closely approximate the real-time measured signal strength values. The coordinates of that entry are considered to be the location of the target station. Better accuracy may be achieved by finding a group of database vectors closely matching the measurements taken from the target transmission. The location estimate is the average of the coordinates in the database entries that were chosen.

An algorithm for comparing real-time target signal strength measurement data and determining the target's location is as follows. An average of a number of readings of the target's signal strength is recorded at each of the APs. The readings are normalized such that the relative received power from the different APs is retained, allowing comparison with the database without regard to different powers radiated by the laptop used in the survey and by the terminal being tracked. A location coordinator that receives the signal strength readings from each of the APs will form a vector $(s_1 \dots s_K)^T$ whose components are the average normalized signal strength of the target at each of K APs in range. This vector must be compared to the database entries according to a given algorithm. One such algorithm, based on minimum Euclidean distance, is described here. For each entry in the database a value D_n is calculated:

$$D_n = \sqrt{\sum_{i=1}^{K} (S_{Ti} - S_{i,n})^2} \qquad (6.14)$$

or more concisely in vector notation as

$$D_n = |\mathbf{S}_T - \mathbf{S}_n| \qquad (6.15)$$

where $\mathbf{S}_T$ is a user online signal strength reading vector, $\mathbf{S}$ is a database signal strength vector, i is an index of the AP, and n is the index of the reference position. The coordinates of the database vector that gives the lowest D is the estimate of the target position.

In the algorithm chosen above, each coordinate in the database has four reference data points, one for each of the four orientations. When the target terminals are apt to be a mix of different types of devices, or different from that used in the survey, such as laptops, desktops, or notebook computers; a possibly better alternative is to average the signal strengths of the orientations of the test target device during the survey measurements and to include one reference point per location coordinate in the database.

Greater precision of the target location may be achieved by choosing more than one neighboring database locations and averaging their coordinates to get an estimate of the target location. For L nearest neighbors, the location estimate is:

$$x = \frac{1}{L} \cdot \sum_{l=1}^{L} x_l \qquad y = \frac{1}{L} \cdot \sum_{l=1}^{L} y_l \qquad (6.16)$$

Creation of a database by collecting individual measurements from a terminal device that is moved from point to point in the whole measurement area may be very time consuming and expensive, particularly when it must be repeated over different areas. An alternative method is to use a propagation formula such as (6.7). n and d_0 can be estimated by making a series of signal strength measurements at different places in the area and estimating the propagation constants by calculating regression parameters. Working on a building construction layout of the area, partition shadowing can be included. Another way to create the database is to use ray tracing software tools. The ray tracing technique calculates wave attenuation over narrow propagation paths between transmitter and receiver, using a three-dimensional representation of the coverage area to determine multipaths and shadowing. The sum of the waves reaching the receiver over multiple paths is used to estimate complex path loss [5]. Ray tracing is used for both indoor and outdoor propagation prediction.

The formation and use of the database can be visualized by maps of signal strength contours related to each base station (AP). An example of such a map for one access point at one of the four position orientations is shown in Figure 6.5. Each contour line has a constant signal strength as indicated. If a nonrandom deterministic propagation law applied in all directions, the contours would all be circles. However, the effects of reflecting walls, furnishings and partitions distort the contours, obligating empirical data collection, ray tracing, or a combination with the propagation formula.

6.3.2.2 Example of Nearest Neighbor RSS Location Method

The following example shows the steps used to obtain coordinates of a mobile target in a RSS nearest neighbor location system. It is based on a simulation of a WLAN operating in an area of 30 by 50m, shown in Figure 6.6. The operating frequency is in the 2.4-GHz band and survey transmitter power is 80 mW. User

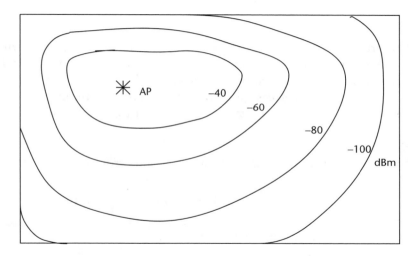

Figure 6.5 Contour map with one AP.

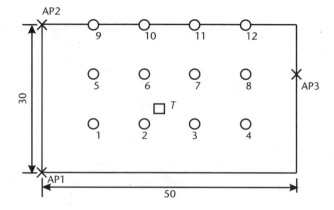

Figure 6.6 Floor plan of WLAN location estimation simulation.

transmitter power is 20 mW. The propagation exponent used in the simulation is 2.5 and reference distance $d_0 = 1$m. A site survey is performed by obtaining average signal strength measurements for transmissions between three access points situated at the positions marked by crosses and reference points having known coordinates located at points marked by circles on the diagram. The data may be taken either by measuring received power from a mobile survey terminal at each AP base station or the power received at the mobile unit from each access point. In this example the mobile survey unit transmits the test signals. The survey unit's antenna is assumed omnidirectional so the recorded average signal strength is for any mobile unit orientation. In a real survey, it may be advised to include in the average at least two different mobile terminal orientations. The survey database showing signal strengths for each survey point and the point's coordinates x, y is shown in Table 6.3. Normalized signal strengths for each survey point are recorded in Table 6.4, where for each survey point the signal strengths relating to each access point are subtracted by the received power at AP1 in dBm.

Example 6.1

The process of estimating the location coordinates of a target terminal is shown in the following steps, along with demonstration measurements. In Figure 6.6, the target position is indicated by a small square. Its true position (x, y) is (23m, 13m). In this example each access point receives and records received signal strength from the mobile transmitter and transmits the data to a location coordinator.

Step 1.
The received power vector at the access points from the target terminal is

$$PT = (-94 \text{ dBm} \quad -96 \text{ dBm} \quad -95 \text{ dBm})^T \tag{6.17}$$

This vector is normalized to

$$PTN = (0 \text{ dBm} \quad -2 \text{ dBm} \quad -1 \text{ dBm})^T \tag{6.18}$$

Table 6.3 Signal Strengths in dBm at Each Reference Location (RL) from Three Access Points

RL	1	2	3	4	5	6	7	8	9	10	11	12
x, y	1, 1	2, 1	3, 1	4, 1	1, 2	2, 2	3, 2	4, 2	1, 3	2, 3	3, 3	4, 3
AP1	−73	−82	−89	−94	−82	−86	−91	−95	−89	−91	−94	−97
AP2	−82	−86	−91	−95	−73	−82	−89	−94	−66	−80	−87	−93
AP3	−94	−89	−82	−73	−93	−87	−80	−66	−94	−89	−82	−73

Table 6.4 Signal Strengths Normalized to the Signal Strength at AP1, for Each Reference Location

RL	1	2	3	4	5	6	7	8	9	10	11	12
AP1	0	0	0	0	0	0	0	0	0	0	0	0
AP2	−9	−4	−2	−1	9	4	2	1	23	11	7	4
AP3	−21	−7	7	−21	−11	−1	11	29	−5	2	12	24

Step 2.

Equation (6.14) or (6.15) is applied to (6.18) and the database in Table 6.4 resulting in a distance metric shown in Table 6.5.

Step 3.

From a search of D (Table 6.5), the survey reference location numbers relating to the three smallest values are found. They are 2, 3, and 6. The coordinates of these locations are found in Table 6.3. Then the average x and y coordinates of these three points are calculated using (6.16), with $L = 3$, to give the estimated location. They are

$$X = 23.33 \qquad Y = 13.33 \tag{6.19}$$

These coordinates give the estimated position of the target. The error is the distance between the estimated and the true position (23, 13). It is 0.47m.

6.3.2.3 Accuracy of Nearest Neighbor RSS

The accuracy of the nearest neighbor RSS method improves as the number of base stations increases. It is affected by the spread of the reference base stations relative to the target according to the geometric dilution of precision (GDOP) [6] (see Chapter 7). Accuracy may also depend on the time of day as the number of people in an office at a given time will affect the actual radiation contours in respect to the database. In one study [7], accuracy in an office environment was given as better than 3m with a probability of 50% and 4.7m for a probability of 75%. The test area in the study had a size of 43 × 25m. There were three base stations operating on the 2.4-GHz band and the database consisted of measurements at 70 survey locations, each with signal strength data from four orientations.

While our example uses an indoor WLAN, similar principles are used for cellular network handset location. Handset location on a college campus is described in [8]. The area covered was 700 × 600m. Three cellular base stations were located in the area and six others outside of it. The data was created using both empirical measurements using a scanner, and propagation estimation. Reported accuracy was 100m 74% of the time and 300m 97% of the time. While this accuracy is not particularly impressive, given the size of the area covered, considering the simplicity and low cost of implementing the RSS method, such results may be suitable for some applications.

6.3.2.4 Bayesian Inference RSS Location Method

Another way of matching a signal strength vector at an unknown location with database reference vectors is by finding the maximum of a likelihood probability

Table 6.5 Signal Strength Vector Differences D Between User on Line Normalized Measured Values and Database Values Per Reference Location (RL) (Figure 6.6)

RL	1	2	3	4	5	6	7	8	9	10	11	12
D	21	6	8	22	15	6	13	30	25	13	16	26

function [9–11]. In the preliminary survey phase, signal strength statistics for the different base stations are determined at reference locations in the coverage area. During the real-time user position estimation phase, the probability of the received signal strength vector is computed for each reference position in the database and the location at which the probability is greatest is the estimate of where the use is situated.

Each state from which signal strength data is collected at the time of the survey is identified as a vector

$$s_k = (x_k \quad y_k \quad p_k)^T \qquad (6.20)$$

where (x_k, y_k) are the coordinates, in two dimensions, of the location and p_k is the orientation of the mobile target terminal. The set of all survey states is

$$S = \{s_1 \quad s_2 \quad \ldots \quad s_k \quad \ldots \quad s_K\} \qquad (6.21)$$

where K, the total number of states, equals the number of survey locations times the number of orientations at each location.

In this discussion we assume that the access points transmit and the mobile station records signal strengths, although the explanation is similar if the mobile unit transmits and signal strengths are noted at the access points. During the database creation phase, a mobile terminal that is similar to the terminals whose locations will need to be estimated polls each access point in turn and makes a number of signal strength measurements from those within range. The measurement sets are taken from each of the position states and identified with the state vector (6.20). Each measurement set is called an observation o.

Observation components consist of a received signal strength measurement and the identification (such as MAC address) of the base station with which the signal strength is associated. In order to discover a probability measure on the signal strength from the access points, each AP transmits multiple messages to the mobile survey unit. The observation at each state is described as follows:

$$o_k = \{(\sigma_1, a_1), (\sigma_2, a_2), \ldots (\sigma_M, a_M)\} \qquad (6.22a)$$

where σ_i is signal strength and a_i is the address of the access point from which the signal strength was measured. M is the number of measurements per state. Each a_i is the address of one of the N access points, b_j, participating in the network, or

$$a_i \in \{b_1, b_2, \ldots, b_N\} \qquad (6.22b)$$

M is generally larger than N, as there are multiple signal strengths reported for each access point. For example, if the first three signal strength readings are from the same access point, AP1, then $a_1 = a_2 = a_3 = b_1$ in (6.22a). All access points may not be in range of all of the states, so some access point signal strengths may be missing for a particular observation.

The Bayesian inference method for comparison of real-time and database statistics is based on Bayes' rule that expresses the relationship between a priori and a posteriori probabilities of form:

$$P(B \mid A) = \frac{P(A \mid B)P(B)}{P(A)} \qquad (6.23)$$

In terms of the position space and observations Bayes' rule can be written:

$$P(s_k \mid o') = \frac{P(o' \mid s_k)P(s_k)}{\displaystyle\sum_{k=1}^{K} P(o' \mid s_k)P(s_k)} \qquad (6.24)$$

where o' is the real-time observation made by the target.

This expression gives the a posteriori probability that the user is in state s_k when he or she makes an observation o'. $P(s_k \mid o')$ is calculated for all s_k and the s_k for which $P(s_k \mid o')$ is maximum is the most likely position state of the target.

$P(o' \mid s_k)$, on the right side of (6.24), is a conditional probability that is found from statistics learned in the survey stage. $P(s_k)$, the a priori probability, is a weighting factor based on the probability distribution of the target over the reference position states S. Generally, the target is just as likely to be at any position in the coverage area, so $P(s_k) = 1/K$ and does not affect the target location estimate. When a moving target is being tracked, after an initial estimation of target position, successive estimates can be made while assigning values to $P(s_k)$, thereby improving the positioning estimation. The denominator of (6.24) is the probability of an observation, $P(O)$ that is independent of position state and therefore does not affect the estimation of s_k. Thus, the probability $P(s_k \mid o')$ is proportional to $P(o' \mid s_k)$, the probability that the user observation o' was made when the target is at position s_k.

The crux of the problem in the Bayesian inference method is to define an expression that can be used to give a numeric value for $P(o' \mid s_k)$ using the statistics derived from the observations that were made during the survey phase. These statistics are derived from observations o_k, (6.22), for each state vector s_k. Two probability relationships expressed as histograms distinguish between the observations made at different reference positions. One of them is the probability of relative frequency of the address of each AP in the set $\{o_k\}$ of observations taken at s_k. The relative frequency equals the number of signal strength measurements in the observation that are associated with access point address b_i (6.22b) divided by the total number of signal strength measurements, M, in that observation. All access points may not be within range of the target at all reference positions, so the frequency distribution of the access points in an observation o_k will give one indication of the reference position s_k. The relative frequency of an AP in an observation is represented by f_i, where $i = 1 \ldots N$, N representing the number of APs. $P(f_i \mid s_k)$ is the probability that the relative frequency of a particularly access point address in an observation o_k equals f_i. The second probability relationship describes the distribution of signal strengths between a target and each access point given s_k, expressed as $P(\sigma_j \mid b, s_k)$. The required conditional probability can then be expressed as [6–8]

$$P(o' \mid s_k) = \prod_{i=1}^{N} P(f_i \mid s_k) \prod_{j=1}^{M} P(\sigma_j \mid b, s_k) \qquad (6.25)$$

where N is the number of access points, M is the number of signal strength measurements, and b is the access point associated with each signal σ_j.

The statistics of $P(f_i \mid s_k)$ and $P(\sigma \mid b_i, s_k)$ are stored as histograms, two for each access point at each state s_k. Figure 6.7 illustrates an example of a pair of histograms. The relative frequency histogram, Figure 6.7(a), shows the possibilities of reception from the associated access point for the states s_k. For example, assume there are five access points and each access point transmits 10 times to the target at each state. The maximum number of signal strengths that can be recorded by the target for that state is $5 \times 10 = 50$. Now assume that access points AP1 and AP2 of the five access points are far from the target at that state and their signals are not always received. Let's say only 3 of the 10 transmissions from AP1 are received, 4 out of 10 from AP2, and all transmissions are received from AP3, AP4, and AP5. Now there are a total of 37 transmissions received for this position state. Relative frequencies for this example are:

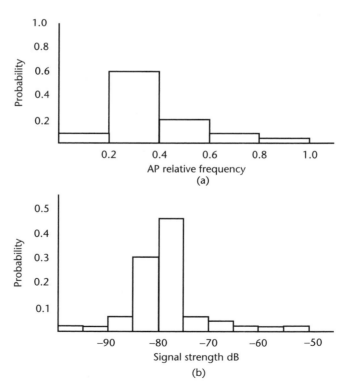

Figure 6.7   (a) Example of histogram for the frequency of a particular access point at one reference position. (b) Example of histogram of the signal strength associated with a particular access point at a reference position.

$$f_1 = 0.08, f_2 = 0.11, f_3 = f_4 = f_5 = 0.27$$

The observations are repeated a number of times, preferably at different times of the day or days of the week, so that there will be an independence of the noise factors that cause repeatedly measured signal strengths from the same access point to differ. Then a spread of relative frequency will be apparent. Figure 6.7(a) indicates that for the access point which the histogram represents, in 60% of the observations, for example, the relative frequency was between 0.2 and 0.4.

Figure 6.7(b) is an example of a spread of signal strengths received from a particular access point for the particular state s_k. Assuming that a total of 200 signal strength readings were made and 92 of them were between −80 and −75 dBm, then the probability of that bin of the histogram is 0.46.

Not all of the relative frequencies or signal strengths in each bin of the Figures 6.7(a) and 6.7(b) will be contained in the observations (6.22) and some estimation of the distribution, or interpolation, will be necessary to give values for all bins. The bin width itself is a parameter that must be determined from the information available in the observation. It is highly desirable that all bins have nonzero values so that (6.25) does not become zero for a particular user observation, so the histograms should be adjusted accordingly [10, 11].

It should be noted that the data in the Bayesian inference method is made up of the statistics of signal strength observations, in the form of histograms, and not of the observations themselves.

During a real-time location estimation, observation measurements o' of signal strength are made, either by the mobile target or by the base stations. No provision is made here for different radiated powers between the mobile terminal used to make the survey and the user mobile terminal, as was done by normalizing the signal strengths to that of one of the access points in the description of the nearest neighbor method above. When the access points transmit to the mobile target, there as no problem, as long as the receiver RSSI calibration is similar to that of the mobile terminal that was used in the survey. In the multilateral case, where the mobile target transmits and signal strengths are measured by the access points, a cooperating mobile unit should include its radiated power in the test messages. The abscissa of the signal strength histogram, Figure 6.7(b), will then need to be corrected by adding the excess (plus or minus) of real-time terminal power in dBm over the power of the test mobile terminal used in the off-line survey.

The use of (6.25) to estimate target location is as follows. After a location estimation command has been initiated in the network, a real-time observation of signal strengths is made, as indicated in (6.22). The location coordinator of the network, or the target, calculates from the observation a relative frequency f_i for each access point. Then the probabilities indicated in (6.25) can be determined from the histograms and $P(o' \mid s_k)$ calculated for each state. The coordinate of the target location is estimated from the coordinates of the states having the highest probability in the a postieri distribution.

6.2.3.5   Comparison Between the Nearest Neighbor and Probabilistic Methods

The database of the Bayesian inference method is larger that that of Euclidian distance and therefore contains more information for use in location estimation.

The latter simply averages all signal readings from each AP where the former includes the details of the probability distribution. In both methods, location estimation accuracy is a function of the number of reference points in a given coverage area, number and placement of base stations or APs, and of course the physical makeup of the coverage area and the effect of the particular environment on radio propagation. The probabilistic method is apt to be better than that of the nearest neighbor when few repeated measurements are made by the user, and when few APs are available in the coverage area [11]. The extent of the data taken during the survey stage affects accuracy of the database components. In the case of the Euclidian distance (nearest neighbor) method, averaging over a large number of measurements will give signal strength results closer to the propagation path mean. In the Bayesian method, a large observation space permits smaller histogram bins and finer probability resolution.

Both methods can rely on the use of propagation laws and ray forming methods to reduce dependence on time-consuming measurements during the survey stage. This entails presumptions about probability distributions as regards the theoretical data for use in the Bayesian method, which may not be accurate.

Reference [10] proposes a Bayesian inference method that yielded a median accuracy of 1.5m as compared to a resolution of 2 to 3m claimed in the study based on Euclidean distance [7]. This advantage of the Bayesian inference method as compared to the Euclidian method cannot be considered conclusive since the experimental trials were not conducted at the same site and under the same conditions.

6.4   Conclusions

Distance measuring and location techniques based on received signal strength have an obvious implementation advantage as compared to TOF methods because they require little or no hardware modification of existing equipment designed for communication. Achievable accuracy is generally less that that obtainable when measuring TOF but is acceptable for many applications. Perhaps the biggest drawback of accurate RSS systems is that they are site/environment dependent, requiring either extensive survey measurements for database creation, or ray tracing analysis using building plans or three dimensional topographic maps. Fingerprinting techniques do have a potential advantage over other techniques since multipath and shadowing modification of propagation are intrinsically included in the database information for a particular site and actually could be a factor in distinguishing between positions over the location system coverage area. In fact, received signal strength data using a fingerprinting algorithm could be supplemented by quantitative multipath information as criteria for location pattern matching.

References

[1]   IEEE Std 802.15.2-2003, "Coexistence of WPAN's," Appendix C.2 Path loss model, August 2003.

[2] IEEE Document 15-04-0461-01-004A for project: IEEE P802.15 "Working Group for Wireless Personal Area Networks," September 2004.

[3] Vig, J., "ISM Band Indoor Wireless Channel Amplitude Characteristics: Path Loss vs. Distance and Amplitude vs. Frequency," Master of Science Thesis, Russ College of Engineering and Technology of Ohio University, June 2004, p. 27.

[4] Abhayawardhana, V. S., et al., "Comparison of Empirical Propagation Path Loss Models for Fixed Wireless Access Systems," *IEEE Vehicular Technology Conference*, Spring 2005.

[5] Zhong, J., et al., "An Improved Ray-Tracing Propagation Model for Predicting Path Loss on Single Floors," *Microwave and Optical Technology Letters*, Vol. 22, No. 1, June 8, 1999.

[6] Wang, Y., X. Jia, and H. K. Lee, "An Indoors Wireless Positioning System Based on Wireless Local Area Network Infrastructure," *SatNav 2003*, Melbourne, Australia, July 2003.

[7] Bahl, P., and V. N. Padmanabhan, "RADAR: An In-Building RF-Based User Location and Tracking System," *IEEE Infocom*, 2000.

[8] Zhu, J., and G. D. Durgin, "Indoor/outdoor Location of Cellular Handsets Based on Received Signal Strength," *Electronics Letters*, Vol. 41, No. 1, January 6, 2005.

[9] Ito, S., and N. Kawaguchi, "Bayesian Based Location Estimation System Using Wireless LAN," *Proc. IEEE 3rd Intl. Conf. on Pervasive Computing and Communications Workshops*, 2005.

[10] Ladd, A. M., et al., "Robotics-Based Location Sensing Using Wireless Ethernet," *MOBICOM*, September 2002.

[11] Roos, T., et al., "A Probabilistic Approach to WLAN User Location Estimation," *International Journal of Wireless Information Networks*, Vol. 9, No. 3, July 2002.

CHAPTER 7

Time of Arrival and Time Difference of Arrival

TOA (time of arrival) and TDOA (time difference of arrival) methods use geometric relationships based on distances or distance differences between a mobile station and a number of fixed terminals to determine the position coordinates of the mobile target. Data for distance estimations are derived from the arrival times of radio signal epochs at one or more receivers. The TOA method uses the transit time between transmitter and receiver directly to find distance, whereas the TDOA method calculates location from the differences of the arrival times measured on pairs of transmission paths between the target and fixed terminals. Both TOA and TDOA are based on the time-of-flight (TOF) principle of distance measurement, where the sensed parameter, time interval, is converted to distance by multiplication by the speed of propagation. In TOA, location estimates are found by determining the points of intersection of circles or spheres whose centers are located at the fixed stations and the radii are estimated distances to the target. TDOA locates the target at intersections of hyperbolas or hyperboloids that are generated with foci at each fixed station of a pair.

Several methods of finding the time of flight have been discussed previously in Chapters 3, 4, and 5. TOA- and TDOA-based location systems may be unilateral or multilateral. In a unilateral system the target communicates with or merely receives fixed terminal transmissions to measure time durations. A multiplexing protocol must be employed since the target must acquire its time data separately from each base station, without those stations interfering with each other. Time, frequency, or code division multiplex techniques may be used. By contrast, a multilateral system performs the location calculations independently of the target, either at one of the base stations or at a separate network infrastructure computing function. The geometric principles of TOA and TDOA location methods are the same for unilateral and multilateral systems.

Ways of applying the two methods to locate a target are first explained while ignoring noise and other impairments. Then the causes of location accuracy deterioration are presented, followed by some algorithms that have been suggested for improving accuracy. Most examples are in two dimensions to simplify presentation and illustration, but extension to three dimensions can be done using the same concepts and methods.

7.1 TOA Location Method

Designers of systems that estimate the distance between two terminals can choose among a number of different system technologies to get position coordinate estimates using TOA. Chapters 3, 4, and 5 detail how time of flight is measured using different communication concepts, and systems employing any of those could provide data for TOA location estimation. Figure 7.1 shows a simple geometric arrangement for determining the location of a target mobile station, MS, that is located on the same plane as base stations BS1 and BS2. The example uses the minimum of two base stations, which for simplification of the calculations are located on the x-axis, with BS1 at the origin. This choice of axes is completely general since coordinates in any rectangular reference frame can be mapped to the arrangement of the figure through coordinate transformation by translation and rotation [1]. The coordinates of BS1 and BS2 are known in advance, and distances d_1 and d_2 are found by multiplying the measured signal propagation time between each base station and the target by the speed of light. It should be noted that data for finding the solution by the TOA method does not have to be obtained from time-of arrival-measurements. The distance readings could be provided by calculating propagation distance directly from transmitter radiated power and received signal strength, when the propagation law is accurately known (Chapter 6). However, due to poor accuracy, distance is rarely measured directly using RSS methods.

The equations for the two intersecting circles with centers at the base stations and radii equal to distances from the target are

$$d_1^2 = x^2 + y^2 \tag{7.1}$$

$$d_2^2 = (x - x_2)^2 + y^2 \tag{7.2}$$

These equations can be solved explicitly for x, y, the coordinates of the mobile station target:

$$x = \frac{d_1^2 - d_2^2 + x_2^2}{2 \cdot x_2} \tag{7.3}$$

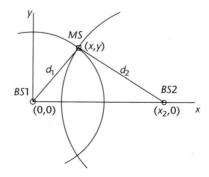

Figure 7.1 Two-dimensional terminal deployment for target location by TOA.

$$y = \pm\sqrt{d_1^2 - x^2} \qquad (7.4)$$

From (7.4) and Figure 7.1 it is evident that y has two possible solutions, one below and one above the x-axis in this example. The true location of the target can be resolved only if there is additional information, aside from the time of arrival data, about where it may be located. For example, it may be known that the target must be in the upper half plane. In this case the negative value for y in (7.4) can be excluded and the target's coordinates are then known, with the value of x given by (7.3).

The TOA method gives the correct location of the target in two dimensions without ambiguity if at least three fixed base stations are used in the measurement. Such an arrangement is shown in Figure 7.2 where an additional base station, BS3, has been added to the base stations BS1 and BS2 of Figure 7.1. The equation of the third circle centered on BS3 and passing through the target location MS is:

$$d_3^2 = (x - x_3)^2 + (y - y_3)^2 \qquad (7.5)$$

Solving (7.1), (7.2), and (7.5) gives the coordinates:

$$x = \frac{x_2^2 + d_1^2 - d_2^2}{2 \cdot x_2} \qquad (7.6)$$

$$y = \frac{x_3^2 + y_3^2 + d_1^2 - d_3^2 - 2 \cdot x \cdot x_3}{2 \cdot y_3} \qquad (7.7)$$

We see that the coordinates of the target can be estimated with no ambiguity since, as seen in Figure 7.2, the position they define is the only one where all three circles intersect.

7.1.1   Overdetermined TOA Equation Solution

Distance measurements that are the basis for TOA location are subject to various causes of imprecision, among them noise, channel interference, multipath, and

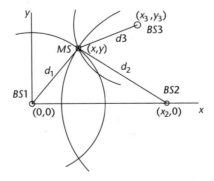

Figure 7.2 Two-dimensional three base terminal deployment for unambiguous target location.

imprecise clocks. Positioning accuracy can be improved by incorporating in the location process a larger number of fixed stations than the minimum required for unambiguous location estimation. Figure 7.3 depicts a two-dimensional layout of four fixed terminals labeled P1, P2, P3, and P4, with known coordinates, and a target terminal P0 whose location is to be determined. If the true distances d_1 through d_4 could be measured exactly, the coordinates of P0 would be at the point of intersection of the circles formed with the fixed stations at the centers and the radii equal to the distances to the target. However, the actual distance measurements, designated D_1, D_2, D_3, and D_4, are not exact, the circles do not cross at one point, and it is necessary to define a criterion for deciding on the estimated location coordinates.

The equations of the four circles defined by base station positions P1(x_1, y_1) through P4(x_4, y_4) and measured distances to target P0, D_1 through D_4, are

$$\begin{aligned}
(1) \quad & (x_1 - x)^2 + (y_1 - y)^2 = D_1^2 \\
(2) \quad & (x_2 - x)^2 + (y_2 - y)^2 = D_2^2 \\
(3) \quad & (x_3 - x)^2 + (y_3 - y)^2 = D_3^2 \\
(4) \quad & (x_4 - x)^2 + (y_4 - y)^2 = D_4^2
\end{aligned} \qquad (7.8)$$

We describe here a method of estimating the position of P0 using a least squares (LS) error criterion. Using the least squares method, the position estimate has coordinates x_e, y_e that minimize the function F:

$$F = \sum_{i=1}^{M} \left(\sqrt{(x_i - x_e)^2 + (y_i - y_e)^2} - D_i \right)^2 \qquad (7.9)$$

where in our example $M = 4$.

Coordinates x_e, y_e that minimize the nonlinear expression (7.9) can be found by an iterative algorithm based on a Taylor series expansion or gradient descent [2, 3]. Such a method may be time consuming and inconvenient to implement for many applications. An alternative approach, described below, gives a closed form solution to the estimation problem. It works by first creating a set of linear equations from the equation set (7.8)

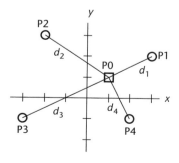

Figure 7.3   Deployment of terminals for overdetermined TOA location.

Expand the factors on the left side of the equations of (7.8) and subtract equations (2), (3), and (4) from (1), to give the following new set of $M-1$, or in this case three, equations:

(1) $\quad (x_1 - x_2)x + (y_1 - y_2)y_e = \frac{1}{2}\left(x_1^2 - x_2^2 + y_1^2 - y_2^2 + D_2^2 - D_1^2\right)$

(2) $\quad (x_1 - x_3)x + (y_1 - y_3)y_e = \frac{1}{2}\left(x_1^2 - x_3^2 + y_1^2 - y_3^2 + D_3^2 - D_1^2\right)$ (7.10)

(3) $\quad (x_1 - x_4)x + (y_1 - y_4)y_e = \frac{1}{2}\left(x_1^2 - x_4^2 + y_1^2 - y_4^2 + D_4^2 - D_1^2\right)$

Equation (7.10) is an overdetermined set of linear equations in x, y. It can be expressed in matrix form as:

$$A \cdot P_e = b \tag{7.11}$$

where

$$A = \begin{bmatrix} x_1 - x_2 & y_1 - y_2 \\ x_1 - x_3 & y_1 - y_3 \\ x_1 - x_4 & y_1 - y_4 \end{bmatrix} \tag{7.12}$$

$$b = \frac{1}{2} \cdot \begin{bmatrix} x_1^2 - x_2^2 + y_1^2 - y_2^2 + D_2^2 - D_1 \\ x_1^2 - x_3^2 + y_1^2 - y_3^2 + D_3^2 - D_1 \\ x_1^2 - x_4^2 + y_1^2 - y_4^2 + D_4^2 - D_1 \end{bmatrix} \tag{7.13}$$

$$P_e = \begin{bmatrix} x_e \\ y_e \end{bmatrix} \tag{7.14}$$

The closed form LS solution to (7.10) is [4–6]:

$$P_e = \begin{bmatrix} x_e \\ y_e \end{bmatrix} = (A^T \cdot A)^{-1} \cdot A^T \cdot b \tag{7.15}$$

While this development uses four base stations, it can be extended logically to a larger number and also to three dimensions, when the number of base stations, M, must be equal to four or more.

The following example demonstrates how (7.15) is used.

Example 7.1

Four fixed base stations at P1 through P4 and a target at P0 are deployed as shown in Figure 7.3. Measured distances between base stations and target are:

Measured distances: $D_1 = 2.5$ $D_2 = 3.2$ $D_3 = 4.8$ $D_4 = 2.5$

The coordinates of the base stations are:

Base station coordinates: P1 = (3, 2)   P2 = (−2, 3)   P3 = (−3, −1)   P4 = (2, −1)

Find a least squares estimate of target position P0.

1. A and b according to (7.12) and (7.13):

$$A = \begin{bmatrix} 5 & -1 \\ 6 & 3 \\ 1 & 3 \end{bmatrix} \quad b = \begin{bmatrix} 1.995 \\ 9.895 \\ 4 \end{bmatrix}$$

2. Substituting in (7.15) the estimated location coordinates are:

$$x_e = 0.823 \quad y_e = 1.396$$

3. The true coordinates of P0 in Figure 7.3 are (1,1). Figure 7.4 shows the circles drawn using the measured distances D_1 through D_4 and the true and estimated target locations.

7.1.2   TOA Method in GPS Positioning

Probably the most widespread system based on TOA is GPS. GPS is a three-dimensional location problem. In order to find three position coordinates, a GPS

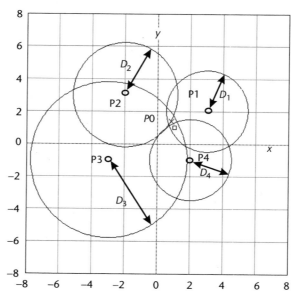

Figure 7.4   TOA circles in overdetermined solution of location equations. X marks the estimated location and the small square marks the true location.

receiver needs to measure the distance to at least three satellites that serve as reference stations, each of whose position in space at the time of epoch transmission is known or can be calculated. The GPS terminal receives satellite data messages that specify the time of transmission of a known epoch of the signal as well as information that the receiver uses to track satellite position. If the receiver had an accurate real-time clock, it could record the time of arrival of the reference signal epoch, then subtract the signal transmission time to get time of flight. The time of flight multiplied by the signal propagation speed is the distance to that particular satellite at the time the ranging message was transmitted. In systems discussed in Chapters 3, 4, and 5 for finding two-way distance between two terminals, it was necessary for each terminal to transmit and receive, one terminal acting as an initiator and the other as a responder, in order to calculate the two-way propagation time between them. In GPS, only one-way transmissions are possible, from the satellite to the receiver. The receiver clock is not accurate enough to use its time-of-arrival measurement to find propagation time by subtracting the satellite's epoch transmission time. Instead, an initial distance calculation, based on the receiver's time-of-arrival clock reading, is made. This distance is called pseudorange. The deviation of the pseudorange from the actual range is the same for all satellites because the same receiver clock is used to make all time measurements, and this deviation can be determined if the pseudorange to at least four satellites is measured.

The following description of how position coordinates may be obtained from pseudorange measurements is based on GPS. The same principles can be applied to a wholly earthbound system, either unilateral, as is GPS, or multilateral (multiple fixed station receivers and target transmitter), as long as it has the following characteristics: reference station coordinates are known at the time of reference station transmission or reception, reference station clocks are synchronized, and time of transmission is conveyed from transmitter to receiver.

The TOA equations in three dimensions are those of spheres whose centers are the known locations of the reference stations and the radii are the distances from each reference station to the target. The target is located on the locus of intersection of the spheres. For four reference stations with coordinates x_i, y_i, z_i, $i = 1$ to 4, pseudoranges R_i, clock offset times propagation speed Δ, and unknown target coordinates x, y, z:

$$(R_1 - \Delta)^2 = (x - x_1)^2 + (y - y_1)^2 + (z - z_1)^2$$
$$(R_2 - \Delta)^2 = (x - x_2)^2 + (y - y_2)^2 + (z - z_2)^2 \qquad (7.16)$$
$$(R_3 - \Delta)^2 = (x - x_3)^2 + (y - y_3)^2 + (z - z_3)^2$$
$$(R_4 - \Delta)^2 = (x - x_4)^2 + (y - y_4)^2 + (z - z_4)^2$$

We can get an insight into what to expect from solving these equations by manipulating them so that the unknown parameters, x, y, z, and Δ are expressed explicitly [7]. This is accomplished as follows. Expand all the square terms in each of the equations (7.16). Then write three new equations that are the first expanded equation minus the second, the first minus the third, and the first minus the fourth.

The x^2, y^2, and z^2 terms are eliminated giving three linear equations with four unknowns that can be rearranged as follows:

$$\delta x_2 \cdot x + \delta y_2 \cdot y + \delta z_2 \cdot z = \delta R_2 \cdot \Delta + \frac{\lambda_2}{2}$$

$$\delta x_3 \cdot x + \delta y_3 \cdot y + \delta z_3 \cdot z = \delta R_3 \cdot \Delta + \frac{\lambda_3}{2} \qquad (7.17)$$

$$\delta x_4 \cdot x + \delta y_4 \cdot y + \delta z_4 \cdot z = \delta R_4 \cdot \Delta + \frac{\lambda_4}{2}$$

which contains compacted constants whose values are expressed as:

$$\delta x_i = (x_i - x_1)$$
$$\delta y_i = (y_i - y_1)$$
$$\delta z_i = (z_i - z_1) \qquad (7.18)$$
$$\delta R_i = (R_i - R_1)$$
$$\lambda_i = x_i^2 + y_i^2 + z_i^2 - x_1^2 - y_1^2 - z_1^2 \text{ for } i = 2 \ldots 4$$

Equation (7.17) can be solved for x, y, and z in terms of Δ and the symbolic values of (7.18). The following example problem demonstrates the use of (7.17) and (7.18) to get a numerical solution [7].

Example 7.2

Figure 7.5 is a simplified geometric representation of the Earth and a GPS satellite orbit. The origin is at the center of the Earth. The positive z-axis extends through the North Pole, positive x intercepts the equator at the prime meridian, that is, longitude 0, and the positive y-axis crosses the equator at latitude 90E. The radius of the Earth is r_e and the radius of the satellite orbit is r_s. Any point P can be specified in polar coordinates (r, θ, φ) or rectangular coordinates (x, y, z). In order to simplify the numbers in this example, all distances are normalized by dividing by the radius of the Earth, whose mean value is 6,360 km. Thus the normalized Earth radius $R_e = 1$. The distance to a satellite directly overhead from the target terminal is 20,200 km, and its normalized height = 20,200 km/6,360 km = 3.176. The normalized radius of the satellite orbit $R_s = 3.176 + 1 = 4.176$. Coordinates of four satellites that are derived from data contained in received messages and measured times of flight are shown in Table 7.1. The times of flight, offset by an unknown clock bias, are used to calculate the pseudoranges R_1, R_2, R_3 and R_4.

The values in the second and fourth columns of Table 7.1 can be used to find the constants defined in (7.18) and then substituted in (7.17). The result is

$$-0.288x + 1.073y + 0.932z = 0.078\Delta - 0.325$$

$$-1.17x - 0.819y - 1.256z = 0.134\Delta - 0.559 \qquad (7.19)$$

$$1.423x - 0.124y - 1.256z = 0.134\Delta - 0.559$$

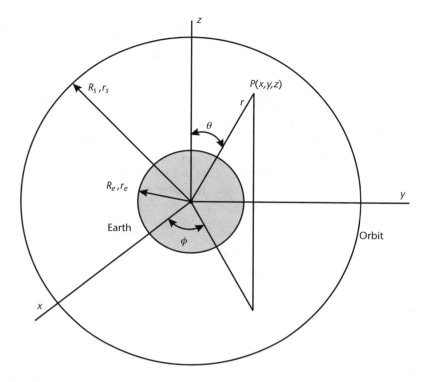

Figure 7.5 Geometric representation of the Earth and a GPS satellite orbit.

Table 7.1 Satellite Data

Satellite (i)	Normalized Position (x_i, y_i, z_i)	Measured Time of Flight (ms)	Normalized Pseudorange R_i
1	(0.828, −3.09, 2.684)	87.378	4.119
2	(0.54, −2.017, 3.617)	89.04	4.197
3	(−0.342, −3.909, 1.428)	90.212	4.252
4	(2.251, −3.215, 1.428)	90.212	4.252

These linear equations can be solved for x, y, and z in terms of the distance bias Δ:

$$x = 0.061\Delta + 0.256$$
$$y = 0.229\Delta - 0.955 \quad (7.20)$$
$$z = -0.199\Delta + 0.829$$

In order to find Δ, (7.20) is substituted in any one of the expressions in (7.16), the first one for example, resulting in a quadratic equation in Δ:

$$\Delta^2 - 11.083\Delta + 9.553 = 0 \quad (7.21)$$

This equation has two possible solutions for the range bias:

$$(1) \qquad \Delta = 0.942 \tag{7.22}$$

$$(2) \qquad \Delta = 10.141$$

These values, substituted for Δ in (7.20), give alternative mathematical solutions to the target position shown in Table 7.2, Also shown are corresponding clock biases in milliseconds, which are the calculated values of ($\Delta \cdot$ 6,360 km/300 km/ ms). The normalized distance of the target from the center of the Earth, column 4, is found from the target coordinates by $\sqrt{x^2 + y^2 + z^2}$.

Solving the ambiguity about the actual target position is easy. Column 4 of Table 7.2 indicates that the target is either on the Earth's surface, Solution 1, or well beyond it. Thus the coordinates calculated from (1) in (7.22) give the true position, which are shown in the third column of Solution 1 in Table 7.2.

The above example has demonstrated that the TOA method with four fixed terminals (the satellites) and a time bias can give a wrong solution for the target coordinates. The mathematical development demonstrated is not used in GPS receivers. They will most likely solve the expressions of (7.16) by an iterative process—substituting an estimate for x, y, and z, and then altering the estimate in steps until a solution is arrived at. Several algorithms are available for converging to the best estimate of target coordinates and receiver clock bias, among them the least-mean-square (LMS) and Newton's method [2]. Most often, more than four satellites are in view at one time, or information is available to reduce the dimensions to be solved for, for example if altitude is known or the clock has been previously adjusted, so that there is no possibility of an ambiguity error in the solution.

7.2 TDOA

Another form of location estimation based on time of flight measurements is TDOA. TOA, as we have seen above, needs some degree of coordination between fixed stations and target in order to determine absolute distances. In two-way distance measurement, initiated by one of two terminals, the second terminal serves as a responder that replies to the initiator after an agreed-upon time interval, or in the case of DSSS, after code synchronization. The time transfer method (Chapter 4) also entails a two-way communication sequence to exchange information on the local clock readings of signal epochs. GPS does not require two-way communication between satellites and receivers. However, the satellite transmitters must send at least their clock reading at the instant of epoch transmission, as well as accurate position information. For many applications, however, target location may be necessary when the target transmitter does not adapt its messages to the expectations

Table 7.2 Possible Solutions to GPS Ranging Example

Solution	Clock Bias (ms)	Normalized Coordinates (x, y, z)	Normalized Distance from the Earth Center
1	20	(0.198, −0.739, 0.642)	1.00
2	215	(−0.366, 1.366, −1.186)	1.85

of the receiver [8]. Clandestine transmitter location, radar location, and electronic warfare in general are some examples. In all of these examples, fixed station reference receivers locate a target transmitter, a multilateral situation. We have already encountered a classical TDOA unilateral system, Loran-C, in Chapter 2. A receiver, which desires to estimate its own location, does not need real-time information from the very low-frequency, low-data-rate fixed transmitters in order to calculate its position. The fact that the TDOA location method can operate with transmitters using their normal communication protocol and with no modification of hardware or software, gives it more applications than TOA, except for GPS. While TDOA transmissions do not need to include a special message for the purpose of the location function, they must have a modulated identity that includes a specific epoch that can be recognized by the receivers. TDOA cannot be used where transmitters emit unmodulated carriers.

Instead of measuring the time of flight of a transmission between two terminals, TDOA measures the difference in the times of flight between a target terminal and a pair of fixed reference terminals. Clock synchronization is required only on one side of the communication link—the side of the fixed terminals [6]. At least one additional fixed terminal is required for TDOA per dimension compared to TOA. A TDOA system needs at least three fixed terminals for a two-dimensional location problem and at least four fixed terminals to estimate three-dimensional coordinates [9].

7.2.1 TDOA Measurement Techniques

The TDOA method is based on the difference of time that radio signals arrive at a receiving terminal from a pair of transmitters. Time difference data can be provided in two ways. First we describe the multilateral case, where signals from a transmitting target are received simultaneously at multiple base stations. One way is to record the time a common signal epoch is received at each terminal, then subtract these times over terminal pairs. This may be done by using the autocorrelation properties of the received signal, as described in Chapter 5. A known replica of the transmitted signal is compared to the received signal in a sliding correlator, or a matched filter output is monitored for the instant of a peak. A second way is to cross correlate the two received signals while adjusting their relative timing until noting the occurrence of a peak. The latter method does not require a known, stored replica of the transmitter signal and is particularly advantageous when the characteristics of the target signal are not known in advance. We describe this method in greater detail.

The signals $r_i(t)$ and $r_j(t)$ from the target received at a pair of base station receivers are:

$$r_i(t) = s(t - td) + n_i(t) \qquad (7.23)$$

$$r_j(t) = \alpha s(t) + n_j(t)$$

$s(t)$ is the transmitter wave, $n_i(t)$ and $n_j(t)$ are noise and interference, α is relative amplitude, and td is the difference in the times of arrival whose value is to be estimated. Each base station stores a sample sequence of the received signal over a concurrent time interval T (the base station clocks are synchronized), whose duration encompasses the signal period and td. An offline cross correlation of $r_i(t)$ and $r_j(t)$ is made at one of the base stations or at a location server in the infrastructure:

$$R_{i,j}(\tau - td) = \frac{1}{2T} \int_0^{2T} r_i(t + \tau - td) \cdot r_j(t)^* \, dt$$

$$R_{i,j}(\tau - td) = \frac{1}{2T} \int_0^{2T} \alpha s(t + \tau - td) \cdot s(t)^* \, dt + \frac{1}{2T} \int_0^{2T} s(t + \tau - td) n_j(t)^* \, dt$$

$$+ \frac{1}{2T} \int_0^{2T} \alpha s(t + \tau) n_j(t)^* \, dt + \frac{1}{2T} \int_0^{2T} n_i(t + \tau) n_j(t)^* \, dt \qquad (7.24)$$

where ( )* signifies the complex conjugate.

The integration over $2T$ is carried out each time τ is incremented. Considering only the first term to the right of the equal sign in (7.24), the cross correlation will have a maximum value when $\tau = td$, thereby giving the estimate of the time difference of arrival for the pair of base stations. The other terms in $R_{i,j}(\tau)$, involving one or both noise signals, reduce the resolution of the cross correlation peak, to a degree that depends on the signal-to-noise ratio of the signals and the correlation between the signals and the noise and between the noise components at each receiver.

Implementation of the cross correlation function will generally be carried out by digital processing involving a summation of product terms. The number of multiplications in the digital processor depends on the sampling rate and the length of the sequences. From the point of view of the processing load, it may be advantages to perform the cross correlation in the frequency domain [10]. Let the frequency spectrum representations of the received signals and noise be:

$$RX_i(f) = F[r_i(t)] \qquad (7.25)$$

$$RX_j(f) = F[r_j(t)] \qquad (7.26)$$

$$S(f) = F[s(t)] \qquad (7.27)$$

$$N_k(f) = F[n_k(t)] \qquad (7.28)$$

where $F[\ ]$ denotes the Fourier transform.

The cross power spectral density $P_{i,j}(f)$ equals the Fourier transform of the cross correlation:

$$P_{i,j}(f) = RX_i \cdot RX^*_j$$

$$= [S(f) \cdot e^{-j\omega \cdot td} + N_i(f)][\alpha S(f)^* + N_j(f)^*] \qquad (7.29)$$

$$= \alpha S(f)^2 e^{-j\omega \cdot td} + S(f)N_j(f)^* e^{-j\omega \cdot td} + \alpha S(f)^* N_i(f) + N_i(f)N_j(S)^*$$

The term $s(t - td) = S(f)e^{-j\omega \cdot td}$ is a result of applying the time delay theorem [see Chapter 5, (5.42)].

For high signal-to-noise ratio and no correlation between the transmitted signal and the noise, (7.29) can be approximated as

$$P_{i,j}(f) = \alpha S(f)^2 e^{-j\omega \cdot td} \qquad (7.30)$$

The inverse transform of the power spectrum is the correlation function. In the time domain:

$$F^{-1}[P_{i,j}(f)] = \alpha R(\tau)^* \delta(\tau - td) = \alpha R_{i,j}(\tau - td) \qquad (7.31)$$

where $F^{-1}[\;]$ is the inverse Fourier transform and $(\;)^*(\;)$ denotes convolution. The rightmost term of (7.31) is a result of the time-shifting property of the delta impulse function $\delta(\;)$ in the convolution. Equation (7.31) shows that after taking the inverse Fourier transform of the cross power spectrum, the time delay td can be found by locating the peak of the resulting cross correlation function. Instead of the processing involved in computing the cross correlation directly as indicated in (7.24), Fourier transform and inverse transform operations are performed, culminating in (7.31). Whereas the number of multiplications to be done in the former method is approximately N^2, where N is the number of sequence samples, in the latter case it is of the order of $N + N \log_2 N$. The frequency domain method may then be preferable when the number of samples per sequence is high. Another advantage in frequency domain processing is greater convenience in neutralizing the distortion caused by the wireless channel response at the base stations [10].

7.2.2   Multilateral and Unilateral Topologies for TDOA

The way time difference measurements are taken in TDOA depends on whether the system is unilateral or multilateral. In a unilateral system the target is a receiver. Assuming the system uses only one frequency, the fixed station transmitters, a minimum of three for two dimensions and four for three dimensions, transmit at different times so that they won't interfere with one another. The clocks of the transmitters are synchronized and the transmitters send periodic beacon transmissions at staggered times so that there is no overlap of transmissions. Figure 7.6 shows a timing diagram of a unilateral system. The target receiver knows the time difference between the epoch transmissions of the transmitters, and can therefore note the time difference of arrival as the interval between the epochs of the received signals as measured by the receiver clock minus the known difference of time between the two transmissions. Figure 7.6 shows the transmission times of three spatially separated transmitters, TX1, TX2, and TX3, and the times of reception

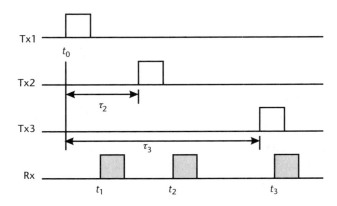

Figure 7.6 Unilateral TDOA timing diagram.

at the target receiver, RX. The receiver knows that TX2 starts its transmission τ_2 seconds after the start of the TX1 transmission at t_0, and TX3 transmits τ_3 seconds after t_0. The clocks of the three transmitters must be synchronized so that each will know when to start its transmission. Another arrangement, that doesn't require clock synchronization, is for fixed terminals TX1 and TX2, which in this case are transceivers, to start their transmission at a fixed time after receiving the signal from TX1 [11]. The times τ_2 and τ_3 then will also include the propagation delays from TX1 to each of the other two terminals. The target receiver clock is not synchronized to the fixed station clocks, so its recorded times of arrival of the signals from the fixed stations, t_1, t_2, and t_3 are relative to any zero reference. The time differences of arrival determined by the target are:

$$\Delta t_{2,1} = (t_2 - t_1) - \tau_2$$
$$\Delta t_{3,1} = (t_3 - t_1) - \tau_3 \qquad (7.32)$$
$$\Delta t_{2,3} = (t_3 - t_2) - (\tau_3 - \tau_2)$$

Two of the three possible time-of-arrival differences, along with knowledge of the fixed station position coordinates, are required to estimate target coordinates in a two-dimensional plane.

The multilateral case is slightly simpler. In the timing diagram of Figure 7.7, target transmissions Tx are monitored simultaneously by the fixed terminals Rx1, Rx2, and Rx3. The clocks of the fixed receiving stations are synchronized. Each of these stations marks the arrival time of the transmission from TX1: t_1, t_2, t_3. The TDOAs are:

$$\Delta t_{2,1} = t_2 - t_1$$
$$\Delta t_{3,1} = t_3 - t_1 \qquad (7.33)$$
$$\Delta t_{3,2} = t_3 - t_2$$

The location computing entity in the system knows the positions of the fixed stations and uses those positions and at least two TDOA values to estimate the

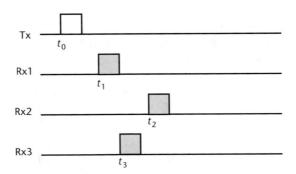

Figure 7.7 Multilateral TDOA timing diagram.

target position. An example of a multilateral TDOA system was given in Chapter 5, Section 5.7.3.

7.2.3 TDOA Geometric Model

The geometric model for estimating position coordinates using time differences of arrival is the intersection of hyperbolas, in two dimensions, and the intersection of hyperboloids in three dimensions. If a base station is located at one focus of a hyperbola and another base station at the other focus, then for a target positioned on the hyperbola the difference between the distances between it and the two base stations is constant. This is shown in Figure 7.8. The hyperbola is characterized by the fact that the difference in distance, $d_2 - d_1$, between any point on it and

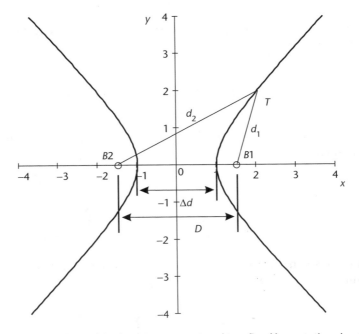

Figure 7.8 Geometric relationships between a target and two fixed base stations in a TDOA system.

the two foci is constant. The difference value is positive if the point is located on the right branch of the hyperbola and negative if it is located on the left branch. The distances are expressed as follows:

$$d_2 = \sqrt{y^2 + \left(x + \frac{D}{2}\right)^2} \tag{7.34}$$

$$d_1 = \sqrt{y^2 + \left(x - \frac{D}{2}\right)^2}$$

The equation of the hyperbola that defines the locus of the target is then

$$\Delta d = d_2 - d_1 = \sqrt{y^2 + \left(x + \frac{D}{2}\right)^2} - \sqrt{y^2 + \left(x - \frac{D}{2}\right)^2} \tag{7.35}$$

where (x, y) are the coordinates of the target, D is the distance between base stations located equal distances from the origin on the x-axis, and Δd is the constant difference of distances from target to base stations. $\Delta d = \Delta t \cdot c$, the time difference of arrival times the speed of light.

Equation (7.35) can be converted to a form that better reveals its identification as a hyperbola [12]. Move the rightmost square root term to the left side of the equation and square both sides:

$$\Delta d^2 + 2 \cdot \Delta d \sqrt{y^2 + \left(x - \frac{D}{2}\right)^2} + y^2 + \left(x - \frac{D}{2}\right)^2 = y^2 + \left(x + \frac{D}{2}\right)^2 \tag{7.36}$$

Move the terms on the left side except the square root term to the right side and square both sides.

$$4 \cdot \Delta d^2 \cdot \left(y^2 + \left(x - \frac{D}{2}\right)^2\right) = (2xD - \Delta d^2)^2 \tag{7.37}$$

After expanding and rearranging the resulting equation, the following expression is obtained:

$$\frac{x^2}{\frac{\Delta d^2}{4}} - \frac{y^2}{\left(\frac{D^2}{4} - \frac{\Delta d^2}{4}\right)} = 1 \tag{7.38}$$

This expression conforms to the standard equation for a hyperbola oriented like the one in Figure 7.8, which is

$$\frac{x^2}{a^2} - \frac{y^2}{b^2} = 1 \tag{7.39}$$

Constants a and b are each one half of the major axis and minor axis, respectively [1]. Equation (7.38) can be translated and rotated to express a hyperbola in any location and orientation.

The location of a target terminal is at the intersection of two or more hyperbolas that are defined from TDOA measurement data. The total number of TDOA values, K, obtainable from M base stations is

$$K = \frac{M!}{2(M-2)!} \tag{7.40}$$

The number of *independent* TDOA values obtainable from M base stations is $M - 1$. All of the independent values in a set are based on at least one measurement of time of arrival between a base station and target that is not used in any other measurement in the set. It is often considered sufficient to include only the independent TDOA in the location estimation process [9]. However, in a noisy environment additional pairs of measurements that are not independent according to the above criterion may be added for redundancy, since the noise that is not correlated between those pairs gives them a degree of independence [12].

An overdetermined system of equations results when $M - 1$ is greater than the number of coordinates, or the dimension, of the desired location. In the presence of noise, the resulting multiple hyperbolas, or hyperboloids, will not intersect at a single point, and criteria must be established for determining the location that provides the best fit to the system of equations. Several methods of estimating location from over determined TDOA measurements are compared in [9, 12].

7.2.4 TDOA Example

The following steps outline a method of finding target coordinates by way of an example based on Figure 7.9. The example equates expressions for the differences of distances between the target and the fixed stations to the measured distance differences. Then it estimates the target coordinates by finding those that give the minimum least square error.

Example 7.3

Base stations at locations P1 through P4 and target at unknown location P0 are deployed as shown in Figure 7.9.

1. Base station and target coordinates
   The known coordinates (x, y) of the four base stations in a local coordinate system are

   $$P1 = (-1.5, -2) \quad P2 = (2, 2) \quad P3 = (-2.5, 2.5) \quad P4 = (2, -1) \tag{7.41}$$

   The true position (unknown to the terminals) of the target is

   $$P0 = (0, 0)$$

   Distances are in kilometers.

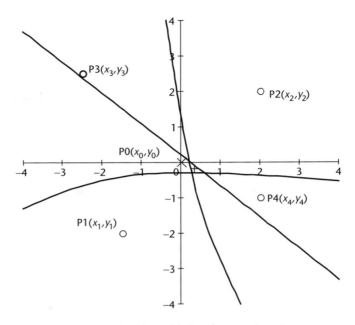

Figure 7.9 TDOA target position estimation with four base stations.

2. Time difference of arrival measurement data and range differences:

   Between P0 and P1, P2: $\tau_{2,1} = 0.095\ \mu s$ $\Delta_{2,1} = 0.03$ km

   Between P0 and P1, P3: $\tau_{3,1} = 4.95\ \mu s$ $\Delta_{3,1} = 1.49$ km   (7.42)

   Between P0 and P1, P4: $\tau_{4,1} = -1.88\ \mu s$ $\Delta_{4,1} = -0.56$ km

   where the range differences $\Delta_{i,j}$ are time differences $\tau_{i,j}$ times the speed of light.

   These measurements include inaccuracies due to noise and interference.

3. Range difference equations: Formulate the expressions for the differences in distance between target and fixed station pairs, that are a function of the unknown target coordinates (x, y) and the known fixed station coordinates.

$$F_{1,2}(x, y) = \sqrt{(y_1 - y)^2 + (x_1 - x)^2} - \sqrt{(y_2 - x)^2 + (x_2 - x)^2}$$

$$F_{1,3}(x, y) = \sqrt{(y_1 - y)^2 + (x_1 - x)^2} - \sqrt{(y_3 - y)^2 + (x_3 - x)^2} \quad (7.43)$$

$$F_{1,4}(x, y) = \sqrt{(y_1 - y)^2 + (x_1 - x)^2} - \sqrt{(y_4 - y)^2 + (x_4 - x)^2}$$

4. Estimate x, y as those values that minimize $F(x, y)$ in the nonlinear least squares expression:

$$F(x, y) = (F_{1,2}(x, y) - \Delta_{2,1})^2 + (F_{1,3}(x, y) - \Delta_{3,1})^2 + (F_{1,4}(x, y) - \Delta_{4,1})^2 \quad (7.44)$$

$(x, y)_{estimate} = (0.32, -0.19)$

In Figure 7.9, the three pairs of hyperbolas do not intersect at the same point. The + symbol marks the least squares estimate and x marks the true target location. The error ΔD is the distance from the + symbol to the true position of the target at P0:

$$\Delta D = \sqrt{(0.32 - 0)^2 + (-0.19 - 0)^2} = 0.37 \text{ km}$$

The solution to Example 7.3 was obtained by minimizing (7.44). Another approach using matrix expressions, similar to that of Example 7.1, linearizes the hyperbolic equation set and solves for a least squares location position with the help of (7.15) [6]. In the case of TDOA, the linear equation set contains, in addition to the unknown position coordinates, the unknown distance between the target and a common reference base station.

Example 7.4

The system is deployed as shown in Figure 7.9 with fixed station coordinates as given in (7.41). Time difference data is the same as (7.42).

1. Define estimated distances between base stations and target.
   Let *D1* through *D4* be distances between base stations at P1 through P4 and target, dependent on measured distance differences of (7.42):

$$D2 = \Delta_{2,1} + D1$$
$$D3 = \Delta_{3,1} + D1 \qquad (7.45)$$
$$D4 = \Delta_{4,1} + D1$$

2. Derive set of linear equations in x_e, y_e and *D1*.
   The system can be represented by a set of equations derived exactly as for the TOA method in (7.10), but with *D2*, *D3*, and *D4* substituted with (7.45):

(1) $(x_1 - x_2)x_e + (y_1 - y_2)y_e = \frac{1}{2}\left(x_1^2 - x_2^2 + y_1^2 - y_2^2 + \Delta_{2,1}^2 + 2\Delta_{2,1}D_1\right)$

(2) $(x_1 - x_3)x_e + (y_1 - y_3)y_e = \frac{1}{2}\left(x_1^2 - x_3^2 + y_1^2 - y_3^2 + \Delta_{3,1}^2 + 2\Delta_{3,1}D_1\right)$ (7.46)

(3) $(x_1 - x_4)x_e + (y_1 - y_4)y_e = \frac{1}{2}\left(x_1^2 - x_4^2 + y_1^2 - y_4^2 + \Delta_{4,1}^2 + 2\Delta_{4,1}D_1\right)$

3. Represent linear equations in matrix form.
   In matrix form, the equivalent equation set is

$$\mathbf{A} \cdot \mathbf{Pe} = \mathbf{b1} + \mathbf{b2} \cdot D1 \qquad (7.47)$$

where

$$A = \begin{bmatrix} x_1 - x_2 & y_1 - y_2 \\ x_1 - x_3 & y_1 - y_3 \\ x_1 - x_4 & y_1 - y_4 \end{bmatrix}$$

$$b1 = \frac{1}{2} \cdot \begin{bmatrix} x_1^2 - x_2^2 + y_1^2 - y_2^2 + \Delta_{2,1}^2 \\ x_1^2 - x_3^2 + y_1^2 - y_3^2 + \Delta_{3,1}^2 \\ x_1^2 - x_4^2 + y_1^2 - y_4^2 + \Delta_{4,1}^2 \end{bmatrix}$$

$$b2 = \begin{bmatrix} \Delta_{2,1} \\ \Delta_{3,1} \\ \Delta_{4,1} \end{bmatrix}$$

$$\mathbf{P_e} = \begin{bmatrix} x_e \\ y_e \end{bmatrix}$$

Similar to (7.15), the estimate of P0 is:

$$\mathbf{P_e} = \begin{bmatrix} x_e \\ y_e \end{bmatrix} = (\mathbf{A}^T \cdot \mathbf{A})^{-1} \cdot \mathbf{A}^T \cdot (\mathbf{b1} + \mathbf{b2} \cdot D1) \qquad (7.48)$$

Making the indicated substitutions in (7.48) we find:

$$x_e = -0.278 + 0.264 D1, \, y_e = 0.414 - 0.26 D1 \qquad (7.49)$$

x_e and y_e are functions of $D1$, which is the radius of a circle whose center is at $P1$ and whose circumference goes through the point $\mathbf{P_e} = (x_e, y_e)$. We find $D1$ by solving

$$(x_e - x_1)^2 + (y_e - y_1)^2 = D1^2$$

After substituting x_e, y_e from (7.49) and x_1, y_1 from (7.47), the result is

$$D1 = 2.582$$

Substituting $D1$ back into (7.49), the estimated position of the target is

$$x_e = 0.404, \, y_e = -0.257$$

This result is reasonably close to the result of Example 7.3, which was obtained by an iterative method of minimizing the least squares.

Several other algorithms have been studied for estimating the target position [9, 12]. There are also various methods for arriving at the minimum values in

(7.44) [2]. The size of the error in the target coordinate estimation relative to the measurement errors depends on the position of the target in relation to the fixed stations. This dilution of precision is covered in more detail in Section 7.3.3.

The methods described in the examples above can be followed logically for two-dimensional systems with more base stations and consequently more equations, and for three dimensional configurations. In the three dimensional case, the distance difference expression for a pair of fixed stations P1 and P2 is

$$F_{1,2}(x, y, z) = \sqrt{(P1_x - x)^2 + (P1_y - y)^2 + (P1_z - z)^2} \qquad (7.50)$$
$$- \sqrt{(P2_x - x)^2 + (P2_y - y)^2 + (P2_z - z)^2}$$

Four base stations, allowing a minimum of three distance difference measurements, are sufficient for three-dimensional location.

7.3   Performance Impairment

There are several reasons why distance and location estimates differ from their true values. Performance impairments are generally common to TOA and TDOA time-of-flight methods, which are both based on time of arrival measurements.

7.3.1   Uncertainties in Data Measurement

Fixed station positions are not known exactly. An example is GPS. The "fixed" stations are satellites that are actually in motion and calculation of their positions is a function of their velocity and various other factors that change with time. These factors are accounted for in a GPS receiver. When fixed station locations are actually static, an improvement in their coordinates may be obtained by interpolating results from trial measurements taken with targets whose locations are known precisely.

Clocks used in the measurement may not be accurate or precise. Clock jitter adds noise to the time measurement and relative drift reduces precision and resolution of the correlation peak, to an extent that is dependent on the correlation integration period. Loss of clock synchronization among reference base stations in TDOA can be a source of inaccurate position results.

Lack of precise knowledge of propagation time is also a source of error. In converting TOA or TDOA measurements to distance, a multiplier of the speed of propagation is used, generally the speed of light. However, the propagation speed is reduced when waves pass through the atmosphere or other forms of matter. In the case of satellite communication, used in GPS, the small reduction of propagation speed as an electromagnetic wave passes through the ionosphere and troposphere must be accounted for in order to improve accuracy. In indoor location systems, building materials through which the signals pass must be accounted for when the location accuracy is in the range of centimeters.

The movement of the target or the fixed stations may affect location accuracy. On the one hand, the Doppler effect may cause trouble in locking transmitter and receiver frequencies, on the other hand, measurement of Doppler drift can be used to correct measurements and also to measure relative movement and speed of the target. Accuracy is affected by the duration of the measuring process when the target is in motion because of the change in relative target position from the start of the measurement to its completion.

7.3.2 Random Noise

There is always noise in a measurement. Receiver noise cannot be totally avoided, and its effect on location precision is a function of the signal-to-noise ratio of the received signal. The effect of noise on the accuracy of the location estimation can be reduced by making multiple time-of-arrival measurements over repeated transmissions and averaging the results before using them in the data processing stage. The individual measurements are spaced in time enough to make them essentially independent in order to take full advantage of the averaging in reducing the measurement variance. The time between readings must take into account momentary interference and fast fading [11]. If the averaging period is long and the target is in motion, the advantage of averaging may be diminished due to position uncertainty caused by a change of location during the measurement. When possible, using a greater number of base stations than the minimum necessary for TOA and TDOA location increases accuracy when least square techniques are employed to solve over determined simultaneous equations [9].

A lower bound to the accuracy that can be achieved in estimating a location is provided by the Cramer-Rao inequality. The Cramer-Rao lower bound (CRLB) takes into account the positions of the base stations relative to the target and the covariance matrix of the TDOA measurements whose components express the degree of independence of the measurements over all transmitter-receiver pairs. A discussion of the derivation of CRLB for TDOA is given in [9]. In the case of TOA, a lower bound for the variance of a time of flight measurement is [13]:

$$\sigma_{TOF}^2 = \left(8\pi^2 \cdot SNR \cdot \sqrt{\alpha} \cdot BW^2 \cdot N\right)^{-1} \qquad (7.51)$$

where SNR is the average signal-to-noise ratio at the two terminals, α is the number of code copies that are averaged, BW is the occupied bandwidth, and N is the number of chips in the code.

7.3.3 Dilution of Precision (DOP)

The effect of range or range difference measurement uncertainties and noise on the accuracy of the location estimation depends on the dispersion of the reference stations in relation to the target. When the reference stations are grouped together, the times of arrival or time differences of arrival will not differ enough to give accurate solutions in solving the simultaneous equations. This is the case when, for example, satellites used for a GPS measurement are all in the same portion of the sky. The best arrangement of fixed stations relative to the target is when the circles, in the case of TOA, or hyperbolas in TDOA, cross at or close to right angles at the target. Figure 7.10 provides a visualization of DOP in two dimensions. In Figure 7.10(a), the base stations are well spread apart. Arcs show the boundaries of the deviation of the range measurements, and the shaded area indicates the uncertainty of the position coordinates. In Figure 7.10(b), the range variance is the same as in Figure 7.10(a), but the area of uncertainty is significantly greater. The example can be extended to three dimensions, where the DOP is equivalent to a ratio of volumes.

Geometric DOP (GDOP) is a unit less number that expresses the amplification of the range errors to the errors in location, and in the case of GPS, time. It is the ratio of the square root of the sums of the variances of the location and clock bias (in distance) errors to the standard deviation of the range error [14]:

$$GDOP = \frac{\sqrt{s_x^2 + s_y^2 + s_z^2 + (c \cdot s_t)^2}}{s_R} \tag{7.52}$$

where s_x, s_y, and s_z are the standard deviations of the x, y, and z coordinate estimations, s_t is the standard deviation of the clock bias estimate, c is the speed of light, and s_R is the standard deviation of the range measurement.

Often the GDOP is expressed as the numerator of (7.52) assuming the standard deviation of the range errors to each fixed station is 1m. Variations of GDOP can be expressed in this manner. A two-dimensional horizontal DOP may be written as

$$HDOP = \sqrt{s_x^2 + s_y^2} \tag{7.53}$$

The dimension of HDOP is unit less, as the coordinate standard deviations are normalized by a 1-m range standard deviation. Other DOP factors are defined as:

- Position dilution of precision, $PDOP = \left(s_x^2 + s_y^2 + s_z^2\right)^{1/2}$

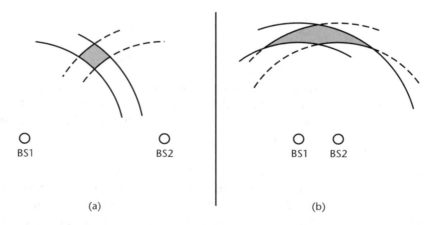

(a)   (b)

Figure 7.10   A geometric illustration of dilution of precision (DOP). The solid curves show the range limits of measurements at BS1 and the dotted lines those at BS2. The target position estimate is within the shaded areas: (a) BS1 and BS2 at optimum separation, and (b) BS1 close to BS2.

- Vertical dilution of precision, $\text{VDOP} = \left(s_z^2\right)^{1/2}$
- Time dilution of precision, $\text{TDOP} = \left(s_t^2\right)^{1/2}$

The best, or optimum, distribution of reference stations relative to the target results in DOP = 1. For GPS, when a defined error budget is maintained for all factors that contribute to the error in the estimated position, DOP will generally be within the limits of PDOP < 6, HDOP < 4, VDOP < 4.5, and TDOP < 2 [14].

More base stations may be available than the minimum required for a position calculation, in which case the DOP can be measured by the location determining entity and an optimum constellation of stations can be chosen for the final position estimation. DOP quantifies a geometric configuration, as shown in Figure 7.10, and all that is needed to calculate it are the coordinates of the reference stations and an approximation of the location of the target. In the case of GPS, PDOP may be interpreted as being proportional to the reciprocal of the volume of a tetrahedron bounded by planes that that are determined by the line-of-sight paths from each of four satellites to the target.

7.3.4 Multipath

The multipath phenomenon is probably the most serious cause of distance measurement and location errors, particularly in indoor applications [13, 15]. Obviously, if TOA or TDOA are measured in nonline-of-sight paths, the distances involved will be greater than the true line-of-sight distances. Usually there are many paths over which the transmitted signal is propagated. If the receiver can measure the time of propagation for each path, it will be able to recognize the shortest path, which is the true distance between transmitter and receiver, unless the line of sight is totally obscured. High bandwidth signals potentially give better discrimination of multipath returns. In order to distinguish between arrival times over two different paths, the minimum receiver bandwidth must be approximately:

$$BW_{mp} \approx \frac{1}{|\tau_1 - \tau_2|} \tag{7.54}$$

where the two times of arrival are τ_1 and τ_2.

There are several methods available for improving a distance measuring or location estimate in the face of multipath propagation [16]. The difference in effectiveness of these methods depends on the strength of the direct signal in relation to the signals that arrive over longer paths, and the receiver bandwidth. We can classify the multipath environment in three categories, labeled DDP, NDDP, and UDP. On a dominant direct path channel (DDP), a predominant portion of the signal energy is received over the shortest line-of-sight path between transmitter and receiver. A nondominant direct path channel (NDDP) contains a direct path with inferior strength as compared to signals reflected from objects outside of the direct path. This case is caused by a physical obstruction on the direct path between the two terminals. For channels in the third category, undetected direct path (UDP), the direct path obstacle is completely opaque to the signal.

Reference [16] compares the performance of three detection schemes in resolving TOA of multipath signals over channels in the three multipath categories. All of them are based on measurements in the frequency domain. In two of the schemes, denoted IFT and DSSS, spectrum samples are converted to the time domain by inverse Fourier transform. A peak detection algorithm estimates the time of arrival. The characteristics of DSSS distance measurement were explained in Chapter 3 from the perspective of a time domain analysis. A simple IFT approach is described in Section 7.2.1. Its ability to distinguish direct path TOA from multipath is conditioned on (7.54). Reference [16] describes a super resolution algorithm called the eigenvector forward backward correlation matrix (EV/FBCM) that significantly improves performance at bandwidths inferior to that specified in (7.54).

The degree of improvement in TOA accuracy that is obtained using the relatively complex EV/FBCM algorithm depends on the category of the multiplex channel and the bandwidth. Under conditions of line of sight (DDP category), above a threshold bandwidth all three schemes perform essentially the same in regard to the mean value of the ranging error. However, when the signal over the line-of-sight path is reduced compared to multipath (NDDP) EV/FBCM performs significantly better than IFT and somewhat better than DSSS. In the third category, UDP, where there is no direct path signal, the EV/FBCM and DSSS estimates have a greater error than for NDDP, but are still better than that of the standard processing scheme, IFT.

Reference [15] refers to several super resolution techniques and describes in some detail a matrix pencil (MP) algorithm for TOA estimation. The authors claim high resolution for their method while using reduced computation complexity as compared to the estimation of signal parameters via rotational invariance techniques (ESPRIT) and multiple signal classification (MUSIC). Reference [15] presents performance measurements and range prediction results that are valid for indoor IEEE 802.11 WLAN systems operating in frequency bands of 2.4 to 2.4835 GHz, 5.15 to 5.35 GHz and 5.725 to 5.825 GHz. Data was taken using a frequency domain measurement set based on a vector network analyzer and the computations for the simulations were carried out on a personal computer. Error performance was shown to be a function of the number of points measured in the complex frequency response, which is proportional to the sampling frequency. For example, a TOA estimation error of 0.05% was noted using 140 sampled points.

Parameter estimation schemes for arrival time and amplitude that operate in the time domain are described in [6]. The interest of this source is location methods for CDMA cellular networks. Basic building blocks of DSSS receivers, including channel searchers and rake correlation processors are combined efficiently in a single architecture in order to detect the individual multipath signal arrivals and then combine them for maximum signal to noise ratio in bit decoding. The rake processor is described pictorially as having fingers, like a garden rake, each of which applies a different incremental shift of the DSSS reference sequence to the multiplier block of one of several correlators. The channel searcher obtains coarse estimates of the strongest multipath signals in order to provide the initial reference signal delay times for use by the rake fingers. Best multipath resolution equals the chip period.

7.3.5   Cochannel Interference

The use of the same wireless channel by several uncoordinated users can cause performance deterioration for communication and for signal time-of-arrival estimation. In the case of CDMA networks, simultaneous transmission on the same channel is a built in operational characteristic of the system. Satellite navigation networks, GPS for example, are based on the cochannel use of all reference transmitters (in satellites) and the estimation of signal arrival times from four or more satellites at a user receiver at the same time.

CDMA cellular networks achieve enhanced cell capacity compared to networks based on frequency division and time division multiple access technologies. However, the key to reduced mutual interference on shared CDMA channels is the maintenance of equal received power at the receiving terminals. Due to the near-far effect, the power output of mobile stations communicating with a common base station must be regulated by dynamic feedback so that the base station receiver will receive the same power from all mobiles in the cell regardless of their distance. When a subscriber location service uses a network-based TOA or TDOA technology, base stations in adjacent cells must estimate times of arrival of signals from a target mobile unit. The mobile can be associated with only one base station at a time, which regulates the mobile's power level, in which case its reception at adjacent base stations will be subject to multiple access interference from transmitting mobiles in that base station's cell.

There are several solutions to the near-far problem as it affects subscriber location. One is by making provision in the network protocol for the mobile station to use maximum power briefly during location message transmissions so that it may be heard by multiple base stations within range. Another is to initiate soft handoff with two or more base stations for the purpose of facilitating simultaneous TOA measurements. Several algorithms have been proposed for delay estimators that are resistant to the near-far effect. They can provide TOA estimates for location in the presence of multiple access interference [3].

7.4   Conclusions

TOA and TDOA techniques have the potential of being the most accurate means of wireless location when they are based on measurements of signal propagation time between communicating terminals (time-of-flight methods). The geometric interpretation of the TOA technique is the estimation of target coordinates by calculating the intersection of circles (two dimensions) or spheres (three dimensions) whose centers lie on the locations of reference base stations and whose radii are the estimated distances between each base station and the target. The same procedure of determining location by geometric calculations holds also when the lengths of the radii are obtained from signal strength measurements, but the discussion on performance impairment in this chapter is in reference to range and location derived from time-of-flight data.

TOA and TDOA techniques may use similar methods for measuring signal time of arrival, but differ in the use of these measurements. In TOA time measurements are converted directly to distances whereas TDOA obtains from them distance differences. As a result, the geometric interpretation of TDOA equations is the intersection of hyperbolas or hyperboloids. However, even when distances cannot be found directly because of a lack of clock synchronization between base stations and target, by adding a dimension to the equations, location coordinates can still be found from intersections of circles or spheres. This is done in GPS receivers, where the added dimension is time bias—the difference of the receiver clock from GPS time. GPS receiver location could also be calculated by TDOA geometry—creating the added dimension by taking a TOA measurement from at least one more satellite than is needed for TOA—and calculating the intersection of hyperboloids. Over determined equation sets, having a number of equations exceeding the number of unknown coordinates, is composed of TOA or TDOA data that is measured from a redundant number of base stations. The equations can be linearized and solved to give a target location estimate based on minimum least squared error. It appears then, that TOA and TDOA methods can achieve equivalent results from the same data.

In order to achieve the highest accuracy from the methods of location based on time of flight, it is necessary to employ techniques that overcome impairments due to noise, multipath, and cochannel interference. The dispersion of reference stations relative to that of the target, measured by the dilution of precision equation, determines the accuracy of target coordinate estimation as a function of the range estimates. Multipath propagation is often the most serious impediment to accurate TOA and TDOA location. Signal processing techniques in frequency and time domain can alleviate the problem. Improved algorithms as well as use of increased bandwidths, notably UWB, are the answer to high accuracy wireless location systems.

References

[1] Spiegel, M. R., *Mathematical Handbook of Formulas and Tables*, New York: McGraw-Hill, 1968.

[2] Widrow, B., and S. D. Stearns, *Adaptive Signal Processing*, Englewood Cliffs, NJ: Prentice-Hall, 1985.

[3] Caffery, J. J., Jr., and G. L. Stuber, "Overview of Radiolocation in CDMA Cellular Systems," *IEEE Communications Magazine*, April 1998.

[4] Golub, G. H., and C. F. van Loan, *Matrix Computations*, 3rd ed., Baltimore, MD: John Hopkins University Press, 1996.

[5] http://www.stanford.edu/class/ee263/ls.pdf, "Lecture 5, Least Squares," 2006.

[6] Sayed, A. H., and N. R. Yousef, "Wireless Location," in *Wiley Encyclopedia of Telecommunications*, J. Proakis, (ed.), New York: John Wiley & Sons, 2003.

[7] Kalman, D., "An Underdetermined Linear System for GPS," *The College Mathematics Journal*, Vol. 33, No. 5, November 2002.

[8] Capkun, S., and J. P. Hubaux, "Securing Position and Distance Verification in Wireless Networks," Technical Report EPFL/IC/200443, Swiss Federal Institute of Technology, Lausanne, May 2004.

[9] Chan, Y. T., and K. C. Ho, "A Simple and Efficient Estimator for Hyperbolic Location," *IEEE Transactions on Signal Processing*, Vol. 42, No. 8, August 1994.

[10] Moore, P. J., I. A. Glover, and C. H. Peck, "An Impulsive Noise Source Position Locator," Final Report, Radiocommunications Agency Contract AY 3925, University of Bath, February 2002.

[11]   Cheng, X., et al., "TPS: A Time-Based Positioning Scheme for Outdoor Wireless Sensor Networks," *IEEE Infocom 2004*.

[12]   Gustafsson, F., and F. Gunnarsson, "Positioning Using Time-Difference of Arrival Measurements," *Proc. IEEE International Conference on Acoustics, Speech, and Signal Processing*, Hong Kong, 2003.

[13]   Lanzisera, S., D. T. Lin, K. S. J. Pister, "RF Time of Flight Ranging for Wireless Sensor Network Localization," *Workshop on Intelligent Solutions in Embedded Systems (WISES '06)*, Vienna, June 2006.

[14]   *Navstar GPS User Equipment Introduction (Public Release Version)*, DoD Joint Program Office, September 1996.

[15]   Ali, A. A., and A. S. Omar, "Time of Arrival Estimation for WLAN Indoor Positioning Systems Using Matrix Pencil Super Resolution Algorithm," *Proc. 2nd Workshop on Positioning, Navigation and Communication (WPNC '05) and 1st Ultra-Wideband Expert Talk (UET'05)*.

[16]   Alsindi, N., X. Li, and K. Pahlavan, "Performance of TOA Estimation Algorithms in Different Indoor Multipath Conditions," *WCNC 2004*, IEEE Communications Society, 2004.

CHAPTER 8

Angle of Arrival

The angle of arrival (AOA) approach to distance measurement and location is probably the oldest method and easiest to understand and to implement. All that's needed is a directional antenna. AOA methods are the core of direction finding (DF), which has been used for years to locate illegal transmitters, both broadcast and those used for eavesdropping, and for tracking wild animals that are tagged with tiny transmitters. Generally, AOA is not restricted by the problems dictating conditions of use of other location methods. It requires no cooperation from the target, and any type of signal can be used, including CW. It also is used over wide frequency bands and ranges—from HF through microwave and from direct true line-of-sight to long communications distances propagated through the ionosphere.

AOA is a principal component in a radar system. Using radar, only one fixed station is required to determine the location of a target in two or three dimensions. The two methods of AOA and TOF are employed. When using AOA alone, at least two fixed terminals are required, or two separate measurements by a single terminal in motion.

8.1 Triangulation

Location and distance are found in an AOA system by triangulation. An example is shown in Figure 8.1. Two base stations are located on the x-axis of a global coordinate system, separated by a distance D. The angles of arrival at the two base stations are α_1 and α_2. From trigonometry we find the coordinates of the target station, (x, y), to be [1]

$$x = \frac{D \tan(\alpha_2)}{\tan(\alpha_2) - \tan(\alpha_1)} \tag{8.1}$$

$$y = \frac{D \tan(\alpha_1) \tan(\alpha_2)}{\tan(\alpha_2) - \tan(\alpha_1)}$$

The angle of the arriving signal cannot be measured exactly. This is illustrated in Figure 8.2. The uncertainty in the measurement of α_1 and α_2 is $\Delta\alpha_1$ and $\Delta\alpha_2$, respectively. The estimated target coordinates are then contained within the shaded region in Figure 8.2. The size of this region, which indicates the possible error of target location, is a function of the measurement accuracy of the angle, the angles themselves, and the distance of the target from the two base stations. The position

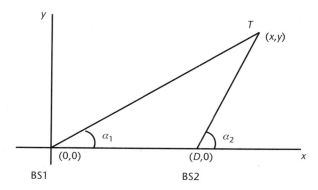

Figure 8.1 Triangulation in two dimensions.

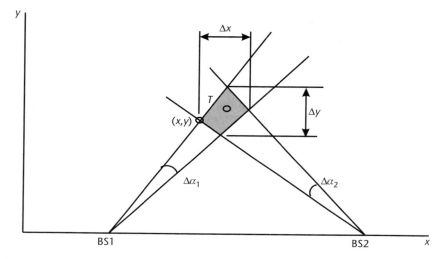

Figure 8.2 Position uncertainty due to antenna beam width.

error is represented by the distance from the estimated location at point $\hat{T}$ with coordinates $(\hat{x}, \hat{y})$ and the true location (x, y):

$$error = \sqrt{(x - \hat{x})^2 + (y - \hat{y})^2}$$ (8.2)

Figure 8.3 shows the variation of the maximum error within the shaded region of Figure 8.2 as the target moves toward BS1 on a straight line having a constant angle α_1. For the curve in the figure, $\alpha_1 = 45°$ and α_2 varies from 70° at the farthest position of the target to 170° at its closest approach to BS1. Beam width of the antennas at BS1 and BS2 is 10°. The scale of the position error in Figure 8.3 is the error distance divided by the distance D between the base stations. Figure 8.3 shows that the position error is maximum when the target is farthest away, at $\alpha_2 = 70°$, and decreases as the target approaches. Minimum error occurs when the bearings from BS1 and BS2 to the target are perpendicular, with α_2 equal to 135°.

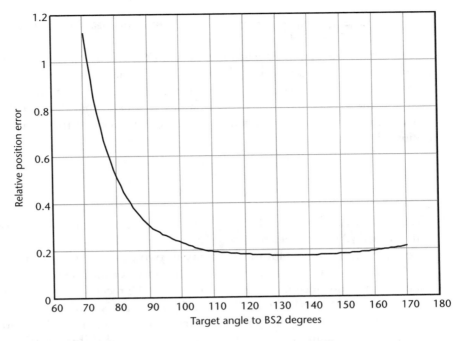

Figure 8.3 Position uncertainty as a function of angle of arrival and range.

It is clear from Figure 8.3 that the position of the fixed stations relative to the target is very important in determining the accuracy of the location estimates. Highest accuracy is achieved when the target and base station positions form an acute triangle. When an included angle between the line connecting the base stations and the line from a base station to a target becomes greater than 90°, the measurement precision is rapidly reduced.

Just as in the other methods of location we have discussed—TOF and RSS—the AOA system can be multilateral or unilateral. In the former, the target transmits and two or more base stations measure angles to the target referenced to a common coordinate system. In a unilateral arrangement, the target takes bearings to fixed base stations whose locations are known. We have seen an example of this in the description of VOR in Chapter 2. In order to find its location relative to two base stations, the target must have an additional point of reference or bearing, such as a compass reading. If the target takes bearings to three base stations whose coordinates are known, it does not need any additional information to find its position relative to those stations. The following discussion assumes that fixed stations are tracking a transmitting target.

8.2 Antenna Performance Terms and Definitions

The description of the AOA method of location revolves around antenna types and characteristics. Terms and concepts used in the discussion of AOA systems are described below. The antenna performance parameters are reciprocal for transmission and reception, so if they are defined in terms of transmission, the meaning for reception is implied.

Antenna directivity expresses the fact that the relative intensity of radiation is different in different directions. Numerically, it is the ratio of the power density at a given distance and direction far from the transmitting antenna (many wavelengths away) to the average power density in all directions at that distance. The average power density is the total radiated power divided by surface area of a sphere whose radius r is the distance:

$$P_{av} = \frac{P_{total}}{4\pi r^2} \tag{8.3}$$

The directivity is a theoretical concept that is a function of the antenna construction and geometry. It may be expressed as a numerical power ratio or in decibels. Antenna patterns are a plot of directivity in two or three dimensions.

Antenna gain is the ratio of the maximum received power from an antenna under test at a distance far from the antenna to the received power from a lossless reference antenna, typically a half-wave dipole or an isotropic antenna, placed at the same height and the same polarization as the tested antenna, when both antennas are fed with the same power. An isotropic antenna is a theoretical antenna that radiates equally in all directions. When the comparison is made with a theoretical isotropic antenna the gain is expressed in decibels in units of dBi, and in units of dBd when the comparison antenna is a half-wave dipole. The isotropic gain is 2.14 dB greater than the gain compared to a dipole. The gain definition is similar to that of the directivity, but it takes into account the tested antenna's ohmic losses. Thus, antenna gain is always less than maximum directivity.

Antenna pattern is a graph that conveniently displays directivity. It is usually presented in polar coordinates. Antenna radiation is in three dimensions but the pattern generally is shown on a plane, usually vertical (elevation pattern) or horizontal (azimuth pattern), although sometimes in the plane that includes the maximum directivity. The appearance of a pattern of given radiation properties depends on the type of scale used to plot the radial component. It may be voltage linear, power linear, logarithmic, or periodically logarithmic. On a logarithmic scale, decibel values are scaled linearly on the plot. The periodically logarithmic scale has been adopted by the organization of radio amateurs, American Radio Relay League (ARRL). It compresses decibel values toward the center of the plot as the radial value becomes smaller. A pattern may be referred to as H-plane pattern or E-plane pattern, depending on whether it is measured in the plane of the magnetic field or the electric field. For a vertical wire antenna, for example, the vertical pattern is the E-plane pattern, since the electric field is parallel to the wire direction.

The radiation *beam width* is the included angle between half power points on the major lobe of the antenna pattern. These are the points that are 3 dB down from the maximum. There is a beam width for the horizontal pattern and for the vertical pattern. The beam width is intimately related to the directivity—an antenna with a narrow beam has high directivity. An approximation of the directivity D expressed as a ratio is obtained from the vertical beam width θ_V and horizontal beam width θ_H using the following formula:

$$D = \frac{4\pi}{\theta_H\,\theta_V \cdot \left(\dfrac{\pi}{180}\right)^2} \tag{8.4}$$

where the angles are in degrees. This approximation is valid for antennas with one major radiating lobe and relatively small minor lobes.

The *polarization* of an antenna refers to the direction of its electric field. Polarization is either linear or elliptic. Linear polarization refers to the case where the electric field lies wholly in one plane containing the direction of propagation [2]. The polarization of transmitting and receiving antennas must be oriented in the same direction for maximum signal strength. In an elliptically polarized wave the extremity of the electric vector describes an ellipse in a plane perpendicular to the direction of propagation, making one complete revolution during one period of the wave. Circular polarization is a case of elliptic polarization where the locus of the tip of the electric vector is a circle. An elliptic polarized emission can be created by feeding perpendicular linear antenna elements 90° out of phase. An axial mode helical antenna has circular polarization for transmission and reception with only one feed point. Its radiation pattern is shown in Figure 8.4. The elliptically polarized wave may rotate in a right-hand or left-hand direction, depending on the positions of the feed points of the linear elements, or on the winding direction

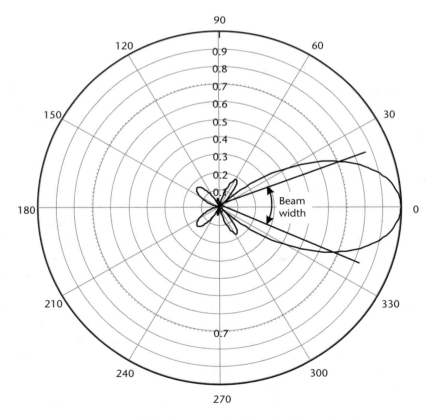

Figure 8.4   Directivity pattern of helical antenna. Radial scale is linear in voltage.

of the wire in the helical antenna. A circular polarized receiving antenna is desired when tracking a moving target with a linear polarized antenna when the direction of polarization varies or is not known.

Directional wire element antenna arrays are often described according to the direction of radiation in respect to the plane of the antenna elements. A *broadside* antenna radiates at right angles to the plane of the elements. Its pattern usually has two main lobes 180° apart. An *end-fire* array has one main lobe in the plane of the parallel elements and perpendicular to them. The Yagi antenna is an example.

8.3   Finding Direction from Antenna Patterns

Antenna directivity is of prime importance in AOA location. A highly directive antenna is needed for good precision. At a given frequency, higher directivity usually entails larger physical size, which is not always possible for many applications. Even an antenna with wide beam width can give accurate direction estimations, if the signal to noise ratio is large. The rms angular accuracy $\Delta\theta$ as a function of beam width θ_B and signal-to-noise ratio SNR is:

$$\Delta\theta = \frac{k\theta_B}{\sqrt{SNR}} \qquad (8.5)$$

where k is a constant of proportionality.

The helical antenna whose pattern is shown in Figure 8.4 [3] can be used for direction finding. The electrical field strength curve in a plane that includes the antenna axis has one main lobe and six minor lobes. The beam width is the angular difference between the points on the curve that are 3 dB down from the maximum, or half-power points. In this case the beam width is 42°. There are two strategies that the terminal may use to estimate the AOA using the signal strength output (RSSI) from a receiver that is connected to the directional antenna. It may rotate the antenna to scan the region of the target at a constant angular rate, then note the antenna angle where the RSSI is maximum. The scan rate must take into account the response time of the receiver signal strength output. It may also scan to locate the target signal, the acquisition mode, then track the target using a servo mechanism type control system to lock the antenna on the target. The signal-to-noise ratio is a dominant factor in determining the AOA. Another factor is the modulation of the target signal. An amplitude varying modulation envelope can make it difficult to determine the peak of the antenna response and it may be necessary to average the signal to smooth out the envelope. Also, a moving target or moving reflecting surfaces near the line-of-sight path of the signal can affect signal strength and impair the accuracy of the AOA estimate.

Minor lobes in the receiver antenna pattern, like those shown in Figure 8.4, may pick up extraneous signals on the same frequency as the desired signal and confuse the direction finding circuits, particularly when those signals are stronger than the desired signal. When the form of the desired signal is known, correlation can be used to single it out from interference.

Instead of finding the peak of the antenna pattern, some direction finding systems use the pattern null to determine the angle of arrival of the target signal. Signal strength changes around the null are much greater than around the peak of the antenna response. Note, for example, the nulls in the pattern of a half-wave dipole antenna, shown in Figure 8.5. The graph shows the received power, scaled in decibels. By tracking the signal null, a simple dipole antenna can be used for direction finding in place of a larger antenna with a more directive pattern. The dipole does have an ambiguity as to which side of the pattern the target is located, but a decision is often possible when there is some knowledge of the target's vicinity. The dipole antenna must be mounted in a horizontal orientation when used for direction finding and would not be useful for vertically polarized targets. A small loop antenna, which can be oriented for vertical polarization, has almost the same pattern as Figure 8.5 and can be used for AOA measurements. For receivers operating in the high frequency band, 3 to 30 MHz, a ferrite coil antenna, whose size is much smaller than a wavelength, has the same pattern as a loop antenna. In order to have a deep null, the dipole or small loop must be very well balanced to its feed line, and be clear of surrounding objects within several wavelengths. The small loop has low efficiency and therefore would not be effective when high gain reception is required.

An antenna having a pattern with a single sharp null and vertical polarization can be made by mounting two vertical dipoles one quarter wave apart in the horizontal direction. When the two elements are fed with currents 90° apart, the pattern is a cardioid, as shown in Figure 8.6. Radial values are on a periodic logarithmic scale. This is an end-fire array antenna, with the maximum gain and null along the line connecting the feed points of the elements, the peak being in the direction of the feed point with lagging phase. Instead of using half-wave dipole

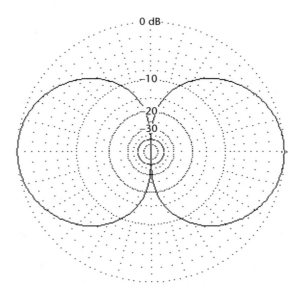

Figure 8.5   Half-wave dipole E-field antenna pattern.

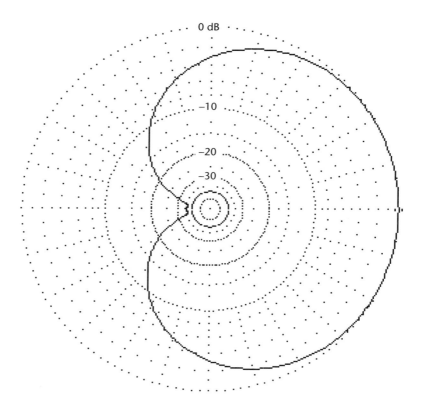

Figure 8.6 Cardiod antenna radiation pattern.

elements, quarter-wave elements can be used instead, with two radials for each, protruding at right angles to the axis of the feed points. Such an antenna array would be particularly suitable for VHF and UHF frequencies. In use, it would be mechanically rotated in search of a signal null.

Two antenna types that have relatively narrow beam widths and are rotated to locate a signal peak are the Yagi array and horn antenna. The Yagi is an end-fire array with parallel linear elements mounted in a plane. In contrast to the phased array of the previous paragraph, there is only one driven element and one or more parasitic elements whose currents are created through electromagnetic coupling. This type of antenna is often used on VHF frequencies, around 150 or 220 MHz, for animal tracking. The dimensions of a five-element Yagi for 220 MHz [4] are shown in Figure 8.7 and its horizontal (azimuth pattern) in Figure 8.8. Beam width is 48° and directivity is 11 dBi. Note that the front to back ratio is 15 dB. A strong interfering signal from a direction opposite the target could make reception difficult.

A horn antenna is practical for AOA on microwave frequencies. It has high gain and a narrow beam width for modest size, and therefore can reject some multipath interference. Such an antenna could be used for indoor applications, inventory tracking in a warehouse, for example, on the unlicensed 2.4- or 5.7-GHz bands. A sketch of a horn antenna for 5.7 GHz is shown in Figure 8.9 [5]. It has

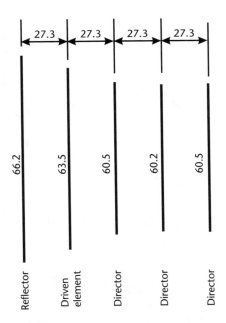

Figure 8.7 Dimensions in centimeters of a five-element Yagi antenna for 220 MHz.

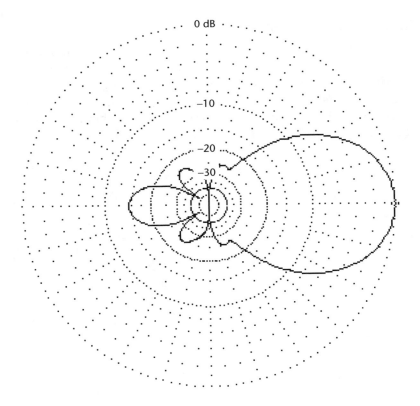

Figure 8.8 Horizontal pattern of 220-MHz five-element Yagi antenna.

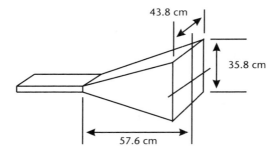

Figure 8.9   Horn antenna for 5.7 GHz.

a directivity of 19 dB. E-plane beam width (vertical pattern for orientation of Figure 8.9) is 16.5° and beam width in the H-plane is 20°.

8.4   Direction-Finding Methods

The outputs of individual antennas in arrays can be combined to obtain angle of arrival with significantly greater precision than that available from the radiation pattern of each element. Two methods are amplitude comparison and phase comparison, implemented as a phase interferometer.

8.4.1   Amplitude Comparison

Instead of rotating a directional antenna, direction-finding can be accomplished by comparing the signal strength outputs of two fixed antennas. The idea is shown in Figure 8.10. Two antennas, each with a cardioid pattern similar to the one in Figure 8.6 but with lower front to back gain ratio are mounted such that their patterns are 180° apart. The receivers have logarithmic amplifiers, so the ratios of the signals from the two antennas are expressed as differences of the outputs, as

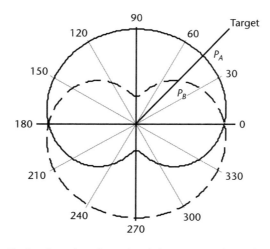

Figure 8.10   Direction-finding by using the ratio of the outputs of two fixed antennas.

shown in the block diagram of Figure 8.11. The log amplifiers and subtraction function can be implemented using a single integrated circuit [6]. The output of the system is plotted in Figure 8.12. Angle of arrival is almost a linear function of the receiving system output, over a range of nearly 180° and 40 dB. Additional information is required to resolve the ambiguity of signals coming from opposite directions. This system greatly increases the direction accuracy obtainable from a not-so-directional antenna. Also, it is not sensitive to common mode interference to the received signals that are canceled in the subtractor after the logarithmic amplifiers. Only the signal-to-noise ratio affects the attainable precision. By employing an array of individual directional antennas, ambiguity can be reduced significantly and precision improved. The coverage angular range will also be reduced so several elemental directional arrays will be required to scan 360°. Antenna pairs are switched in and out to the receivers, depending on the direction of the target.

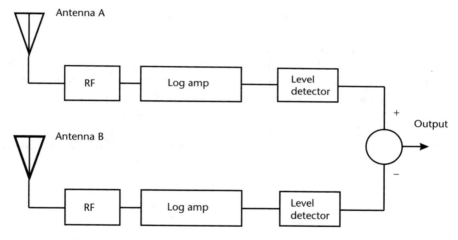

Figure 8.11   180° phase detector receiver.

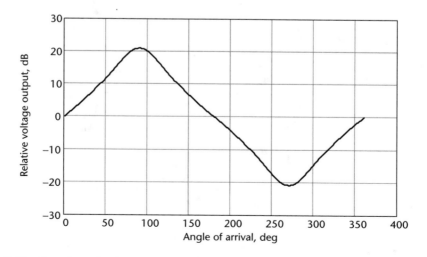

Figure 8.12   Plot of amplitude comparison direction finder output.

Instead of using two antennas with fixed patterns, the offset phase feeds of one antenna array can be switched to give a similar result. One way to do this is to reverse the leading and lagging phase inputs to two antenna elements. Only one receiver is needed for this arrangement. The signal out of the logarithmic amplifier is read and stored during one position of the switch. Then the antenna pattern is switched and a second reading is stored. The stored values are subtracted and used to calculate the angle of arrival, as described above. When the angle of arrival of the target transmission at a distant fixed station is known, location can be calculated.

8.4.2 Phase Interferometer

A direction-finding receiving array can be made using static wide beam width antenna elements. The angle of arrival of a received signal is found by measuring the phase difference between the outputs of the elements. Such a system is called a phase interferometer. Its principle is described with the help of Figure 8.13. When the target is much farther away from the antenna elements than the distance between them, the wave front—locus of points of equal phase—approaching the array is essentially a straight line. For an angle $\theta > 0$, the wave front reaches antenna B before it reaches A, causing a phase difference between the signals at the terminals of the two antennas. The phase difference is measured by a phase detector, whose output is a function of the direction of arrival of the signal from the target. This function is derived as follows.

The signal delay Δt at A with respect to the time the wave front arrives at B is the distance a divided by the speed of propagation c.

$$\Delta t = \frac{a}{c} \tag{8.6}$$

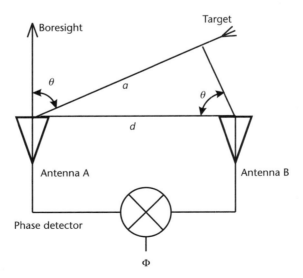

Figure 8.13 Phase interferometer.

a depends on the angle of arrival θ and the distance d between the two antenna elements:

$$a = d \sin(\theta) \tag{8.7}$$

The phase in radians between the signals arriving at A and B is then:

$$\phi = \Delta t \cdot 2\pi \cdot \frac{c}{\lambda}$$

$$\phi = \frac{a}{c} \cdot 2\pi \cdot \frac{c}{\lambda} \tag{8.8}$$

$$\phi = 2\pi \cdot \frac{d}{\lambda} \sin(\theta) = 2\pi \cdot k \cdot \sin(\theta)$$

where

$$k = \frac{d}{\lambda} \tag{8.9}$$

It is apparent from (8.8) that the unambiguous range of θ is $\pm 90°$. The phase interferometer cannot distinguish between targets on one side of the antenna elements from those on the opposite side. Equation (8.8) shows that the span of the phase difference ϕ over the range of θ depends on the constant k. The output of the phase detector, Φ in Figure 8.13, cannot, however, exceed the bounds of $-180°$ and $+180°$, so values of θ that cause ϕ in (8.8) to exceed these bounds cannot be determined unambiguously by reading Φ. The relationship between the angle of arrival and the phase detector output which confines that output to angular bounds can conveniently be written as:

$$\Phi = \arg\left(e^{j \cdot 2\pi \cdot k \cdot \sin(\theta)}\right) = \arctan(2\pi \cdot k \cdot \sin(\theta)) \tag{8.10}$$

When using the phase interferometer of Figure 8.13, it is more convenient to show the angle of arrival θ as the independent variable and Φ as the dependent variable. Their relationship is expressed as

$$\theta = \arcsin\left(\frac{\Phi + n \cdot 2\pi}{2\pi \cdot k}\right) \quad n = 0, \pm 1, \pm 2, \ldots \tag{8.11}$$

The unambiguous range of the angle of arrival that can be measured over the full span of the phase detector output depends on the distance between the two antenna elements relative to wavelength, expressed by k. The maximum range of $-90° < \theta < +90°$ occurs when the argument in parentheses on the right side of (8.11) equals unity when Φ equals π radians and $n = 0$. In this case, $k = 0.5$, for $d = \lambda/2$. A plot of angle of arrival θ versus phase detector output Φ, expressed in (8.11), is drawn in Figure 8.14, for $k = 0.5$. When $k > 0.5$, two or more values of angle of arrival

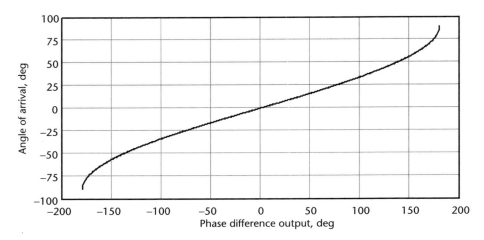

Figure 8.14 Angle of arrival versus phase difference output for $k = 0.5$.

will result in the same value of Φ at the phase detector output. The values of n are chosen such that the argument in (8.11) is not outside ± 1.

The error in the estimate of the angle of arrival as a function of the error in the phase detector output reading is not constant across the coverage range. We find an expression for the error as follows.

$$\Delta\theta = \Delta\phi\frac{d}{d\phi}(\theta(\phi)) \tag{8.12}$$

where $\Delta\theta$ is the error in estimating θ and $\Delta\varphi$ is the error in the phase detector reading as the result of receiver noise or phase error in the receiver chains shown in Figure 8.15. Substituting from (8.11) and taking the derivative:

$$\Delta\theta = \frac{\Delta\phi}{\sqrt{(2\pi \cdot k)^2 - \phi^2}} \tag{8.13}$$

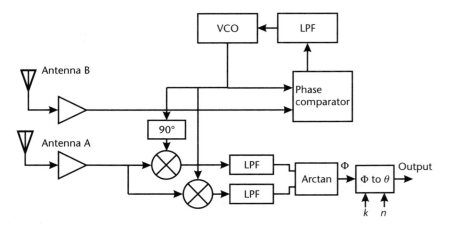

Figure 8.15 360° output phase difference detector.

A plot of the angle of arrival error $\Delta\theta$ versus the angle of arrival θ is shown in Figure 8.16 for a phase detector output error, $\Delta\varphi$, of 1°. It shows the result of (8.13) with φ substituted by (8.8). The plot shows that over an angle of arrival range of ±50°, corresponding to the phase detector output range of −135° to +135° (Figure 8.15), the AOA estimate error is less than 0.5° per phase detector output degree. Beyond this range, the error rises steeply until it approaches infinity as the readout angle approaches the limits of ±90°.

The phase interferometer in Figure 8.15 works as follows. A phase lock loop locks the receiver local oscillator in frequency and phase to the incoming signal at antenna B. A coherent quadrature downconverter provides in-phase and quadrature dc levels which, applied to an arctangent processing function, are interpreted as the phase difference between A and B—the direction of arrival of the signal from the target. The phase difference is then converted to angle of arrival in the Φ to θ block using (8.11) with inputs k and n. The signals must be narrowband, or CW, to give a clear result [7], so that the angle estimation accuracy depends essentially on the SNR of the received signal. The signal from Antenna B cannot be used directly for comparison with the signal on antenna A because of its noise. This noise is reduced by the lowpass filter in the PLL that locks the antenna B signal to the local oscillator.

Considering only receiver noise, we can estimate the angle of arrival error for the implementation of Figure 8.15 using the error plot in Figure 8.16. The RMS phase jitter at the arctangent block output depends on the passbands of the three lowpass filters in the diagram. If we assume that the bandwidth of the phase lock loop filter connected to the phase detector is considerably less than that of the bandwidth in the antenna A receiving chain so that the reference oscillator noise does not significantly affect the phase jitter, the standard deviation of the output angle estimate is approximately $1/\sqrt{2 \cdot SNR}$ (see Chapter 5, Section 5.3). For a $SNR_{dB} = 20$ dB, the standard deviation of the phase jitter is

$$\sigma_\Phi = \frac{1}{\sqrt{2 \cdot 100}} \cdot \frac{180}{\pi} \approx 4°$$

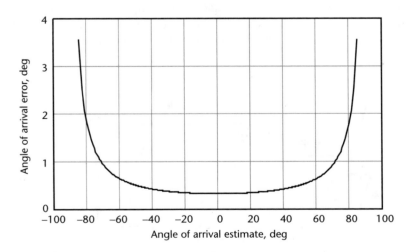

Figure 8.16 AOA estimate error as a function of a 1% error in the phase difference output reading.

Using Figure 8.16 while accounting for the 4° phase jitter, the angle of arrival error in an AOA estimate of 0° is $0.32 \times 4° = 1.28°$ and for an estimate of 80° the error is 8°. The phase interferometer does not give accurate estimates for angles of arrival very close to ±90°; that is, along the axis of the antenna elements.

Another configuration for the phase interferometer is shown in Figure 8.17. It is simpler than that of Figure 8.15 but not as accurate. The outputs of the logarithmic amplifiers are square waves, whose zero crossings reflect the phases of the incoming signals. The phase detector block can be implemented by XOR logic. Its output is filtered and the result is the absolute value of the phase difference between the signals at the outputs of the two antennas. When the angle signals at the input to the phase detector are of equal phase, the output is maximum. The output is zero when the signals differ in phase by 90°. The phase output from the multiplier spans 0° to 180°, as compared to ±180° in the arrangement of Figure 8.17. In order to show plus and minus values of the angle of arrival referenced to boresight of Figure 8.13, a $\lambda/4 = 90°$ delay is inserted in the A receive path. This shifts the 0° phase difference point in the output to a voltage output in the middle of the voltage swing. Then phase differences from −90° to +90° can be read from the circuit. The same integrated circuit, [6], that was referenced for the amplitude comparator of Figure 8.11, has facilities for phase difference measuring as shown in Figure 8.17.

The accuracy of the AOA estimate is improved by increasing the separation between the antenna elements. However, doing so causes ambiguities in relating the AOA to the phase difference output. For example, Figure 8.18 shows the difference output Φ, using the circuit of Figure 8.15, as a function of AOA when $k = 2$, or d equals two wavelengths. The range of the AOA that can be estimated unambiguously from the phase difference is found from (8.11). With $\Phi = 180°$, $k = 2$ and $n = 0$, for example, $\theta_{range} = 2 \cdot \arcsin\left(\dfrac{\pi}{2\pi \cdot 2}\right) = 29°$ centered on $\theta = 0$. When the angle of arrival θ is beyond ±14.5°, an approximation of θ is required in order to specify n in (8.11) which denotes the continuous segment in the plot of Figure 8.18 that resolves the ambiguity. For example, assume it is known that the target is at an angle of between −35° and −45°. The phase difference measurement output Φ is −108°. Choosing $n = -1$, using (8.11) the angle of arrival $\theta = -40.5°$. Note that the angles in (8.11) are in radians.

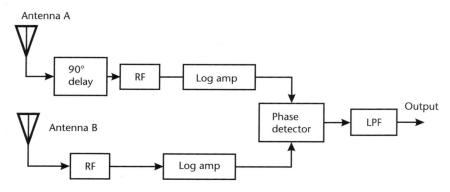

Figure 8.17   A 180° output phase difference detector.

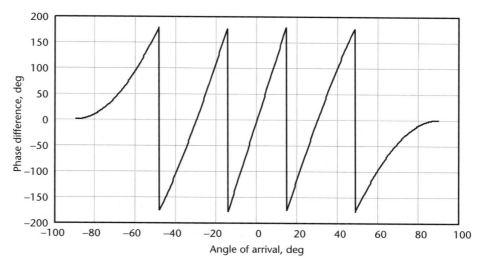

Figure 8.18   Phase difference output versus angle of arrival for antenna element separation of two wavelengths.

While increasing the separation between the antenna elements decreases the unambiguous range, it also increases the accuracy of the angle of arrival estimation. The effect of k (relative element separation) on accuracy is apparent from (8.13). When phase difference $\varphi = 0$, the angle of arrival error is inversely proportional to k. The deviation in the AOA for 1° of error in the phase detector reading is 0.08° for $k = 2$ as compared to 0.32° with $k = 0.5$, when $\varphi = 0$.

One way to solve the ambiguity problem when large element spacing is needed for accuracy is to provide three antenna elements instead of two. The distance between two of the elements is the minimum spacing for maximum coverage, which is $\lambda/2$ for $k = 0.5$. The distance between one of those elements and a third element is as needed for required accuracy, for example two wavelengths ($k = 2$). One of the receiver inputs is switched between two of the antenna elements. When elements with the minimum spacing are connected, the approximate AOA can be estimated, then detection with the maximum spacing will give a higher accuracy estimation of the angle to the target.

8.4.2.1   Implementation Using Multielement Antenna Array

A proposal for using the phase interferometer method for tracking wild animals is an illustration of how the basic principles discussed above may be implemented [8]. The system operates at 144 MHz (wavelength equals 2m) and consists of an array of eight individual vertical antennas, equally spaced by 1m in a straight line. Each antenna consists of two closely spaced parallel elements whose feed points are phased to provide a directional pattern that has a front to back ratio to reduce sensitivity to interfering signals coming from the opposite direction from that of the target bearings. An AOA estimator algorithm called multiple signal classification (MUSIC) was selected. This algorithm provides simultaneous bearings to multiple targets and inherently solves the ambiguity problem while improving the AOA estimation in the presence of multipath. Instead of switching the antenna outputs to two receivers, as was described in our development above, the outputs of all array antennas are sampled to provide a vector for processing by the algorithm.

For implementing the chosen algorithm in [8], a normalized vector composed of the outputs of M antennas in the array, ignoring noise, is written as:

$$\mathbf{a}(\theta) = \left[1, \exp\left(-j\frac{2\pi \cdot d \sin \theta}{\lambda} \right), \ldots, \exp\left(-j\frac{2\pi(M-1)d \sin \theta}{\lambda} \right) \right]^{T}$$

(8.14)

The negative exponents indicate phase delay as compared to the reference component in the first row of the vector. When responses are received from additional sources in the surveyed area, there will be several angles of arrival, $\theta_1, \theta_2, \ldots, \theta_L$. An $M \times L$ array that includes all the vectors of type (8.14) is

$$\mathbf{A} = [\mathbf{a}(\theta_1), \ldots, \mathbf{a}(\theta_L)]$$

(8.15)

The M-dimensional signal vector received from L tracked sources is:

$$\mathbf{x}(t) = \mathbf{A}\mathbf{s}(t) + \mathbf{n}(t)$$

(8.16)

where $\mathbf{s}(t)$ is an L-dimensional vector of the signal sources and $\mathbf{n}(t)$ is the vector of the M noise components on the M antenna outputs [7, 8]. $\mathbf{x}(t)$, then, is the model of the signals from the antenna array that is the input to the MUSIC algorithm. The output is a spectrum plot with abscissa calibrated in azimuth angle. Spectrum peaks are identified as the angle of arrival of the signals from the tracked sources. For tracking the wild animals, arrays as described are placed at positions around the surveyed area. The results of the tracking algorithm from each array subsystem are transmitted to a central processing server that calculates the position of the tracked animals.

8.4.2.2 Influence of Elevation Angle on AOA Accuracy

Note that the angle of arrival, θ, used in the above development refers to angles that are in the plane that is common to the target and the two antennas, which is not necessarily the horizontal plane. When the target is at a high elevation, the angle to the target projected on the horizontal plane differs from θ and corrections may be needed when defining the target coordinates. Equation (8.17) shows the relationship between the projected horizontal angle of arrival, θ_a, and phase detector output φ when the angle of elevation, α, is accounted for:

$$\phi = 2\pi \cdot k \frac{\sin(\theta_a)}{\cos(\alpha)}$$

(8.17)

Figure 8.19 shows the horizontal angle to the target as a function of the phase detector reading for target elevation of 0° and 45°. When the elevation is not

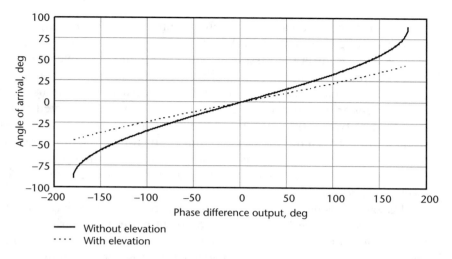

Figure 8.19   Horizontal AOA versus phase difference output with and without compensation for elevation of 45°. Antenna separation parameter $k = 0.5$.

known and is considered zero, the solid line plot is interpreted as the target azimuth. However, using $\alpha = 45°$ in (8.17), the true azimuth is shown by the dotted line. For example, a target that gives a phase detector output of 100° would be thought to have an azimuth of 34° when elevation is not accounted for. The dotted curve shows that the actual azimuth is 23°, a significant variation. When the target is located closer to the boresight of the antenna, and when elevation is lower, there is not a large difference between corrected and uncorrected azimuth readings. For large azimuth deviation from boresight and significant elevation, the span of the angle of arrival is reduced because the limit of the phase difference detector is reached before the AOA achieves ±90°.

The phase interferometer is a relatively simple direction-finding device, considering the antenna structure and electronic circuitry. However, it does have limitations compared to other directionfinding methods, particularly those to be discussed in the following sections. First of all, it applies only to receiving systems. It has maximum unambiguous span of 180° and cannot distinguish signals arriving from front or back. In order to prevent signals arriving from the opposite direction from interfering with the measurement, antenna elements with large front-to-back ratio are used. The fact that these elements have greater directivity than simple omnidirectional elements may limit the useful azimuth range of the direction-finding system. A particular deficiency of the interferometer is that direction-finding is achieved without using highly directive antennas. Thus, it does not have the gain and interference rejection capabilities of other direction-finding methods.

8.5   Electronically Steerable Beam Antennas

Electronically steered antennas and arrays create directive beam patterns whose form and aiming are adjustable by software in order to give high flexibility both for transmitting or receiving in a desired direction, as well as nulling out unwanted signals. This discussion assumes receiving antennas but the principles are applicable to transmitting arrays as well. Adaptive arrays are called smart antennas, which may be divided into two classes: switched beam and steerable adaptive arrays [1].

An example of a simple switched beam antenna is an array of directive antennas arranged on a circle, as shown in Figure 8.20. Adjacent patterns overlap so that all directions are covered. During reception, a target can be located by consecutively switching between the antenna elements while measuring the received signal strength. In between directions are estimated by noting the ratio of signal strengths of adjacent lobes. Cellular base stations usually use three switched antennas to divide the cell into 120° sectors. The switches themselves may be solid state or mechanical relays.

Instead of using antennas with fixed patterns, the radiation from individual antenna elements can be combined to create patterns of various shapes [9]. The antenna elements must by separated in space so that each one sees an independent phase of the received signal at a given time instant as compared to the other elements. By varying the phase of the signal taken from each element, the pattern is adapted to have desired characteristics. There are two directional qualities that can be controlled: direction of maximum signal strength and direction of null. The null is much sharper than the lobe maximum, and is often used for direction finding. Also, controlling null direction allows reducing interfering transmissions on receiving, and limiting interference from a transmitter in a desired direction. Adaptive antenna systems may be open loop or closed loop. An open loop system decides on the desired pattern with no regard to the signals actually being received and applies steering parameters stored in system memory. One set of parameters may be replaced by other sets from memory until the required result is obtained—maximum signal or interference cancellation. Thus, the open loop system can employ feedback, but the parameters that determine the pattern are fixed in advance. A closed loop adaptive system modifies the antenna pattern parameters according to real time desired and undesired signals and noise. The constants that set the phase for each element are modified by signal processing for a given situation that is not stored in advance.

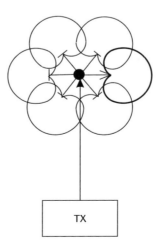

Figure 8.20 Switched directional antenna array.

Figure 8.21 shows a three-element adaptive array. Each element is an omnidirectional antenna—a vertical half-wave dipole or quarter-wave monopole mounted on a ground plane, for example. The desired pattern is in the horizontal plane. The elements A, B, and C are aligned in a straight line, and the target direction is limited to ±90° from boresight. The outputs of elements B and C are each applied to an electronically adjusted phase shift network preferably after RF amplification and down conversion to an intermediate frequency. The phase shift in all channels up to the phase shift networks must be kept equal. The outputs of the phase shifters are summed and the power of the resultant signal is determined. We saw in the section on the phase interferometer how the phase at antenna element B and similarly at C changes with the angle of arrival, θ, of the received signal. Utilizing (8.8):

$$\beta_1(\theta) = 2\pi \cdot \frac{d}{\lambda} \sin \theta \tag{8.18}$$

$$\beta_2(\theta) = 2\pi \cdot 2\frac{d}{\lambda} \sin \theta \tag{8.19}$$

where $\beta_1(\theta)$ and $\beta_2(\theta)$ are the phases at elements B and C, respectively.

If the input signal to antenna A, with normalized amplitude, is $\cos(\omega t)$, the output signal from the summing network is

$$S_{out} = \cos(\omega \cdot t) + \cos(\omega \cdot t - \beta_1(\theta) + \alpha_1) + \cos(\omega \cdot t - \beta_2(\theta) + \alpha_2) \tag{8.20}$$

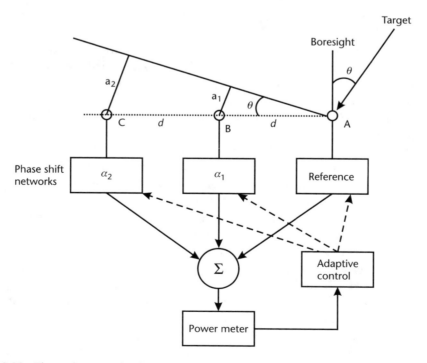

Figure 8.21   Three-element adaptive antenna array.

where ω is the angular carrier frequency and α_1 and α_2 are the settings of the phase shift networks. It is convenient to develop an expression for the signal power output using complex baseband notation. The phase relationships at the outputs of antenna elements A, B, and C as a function of the angle of arrival θ are represented in the steering vector:

$$\mathbf{s}(\theta) = [1 \quad \exp(-j \cdot \beta_1(\theta) \quad \exp(-j \cdot \beta_2(\theta)]^T \qquad (8.21)$$

The vector that shows the setting of the phase shift networks is:

$$\boldsymbol{\alpha} = [1 \quad \exp(j \cdot \alpha_1) \quad \exp(j \cdot \alpha_2)]^T \qquad (8.22)$$

The power at the output of the summing block in Figure 8.21 is concisely expressed as:

$$P_{out} = \left| \frac{\mathbf{s}(\theta)^T \cdot \boldsymbol{\alpha}}{M} \right|^2 \qquad (8.23)$$

where M, a normalizing constant, is the number of array elements, in this case three.

When networks α_1 and α_2 are adjusted to subtract out the phase shifts at B and C relative to A ($\beta_1(\theta)$ and $\beta_2(\theta)$), then all three signals into the adder will be the same and their sum will have three times the amplitude of each signal for maximum power. In a closed loop, the adaptive controller adjusts α_1 and α_2 for maximum power output of a received signal coming from an unknown angle of arrival. The angle of arrival can then be estimated from a mapping table of the resulting values of α_1 and α_2. In open loop operation, prestored sets of values for α_1 and α_2, corresponding to given values of θ, are tried and the θ value corresponding to maximum power is estimated as the AOA. Greater precision can be obtained by interpolating between the adjacent highest power points.

Figure 8.22 shows sample antenna array patterns when the phase networks are set for angle of arrival values of 0°, 15°, −30°, and 45°. The radial scale is linear with power. Antenna elements are separated by one half wavelength: $d = \lambda/2$. Beam width is relatively narrow for small angles of arrival around boresight but increases as the absolute value of θ increases. There is also a significant side-lobe at −90° when θ equals 45°. It is evident from the radiation patterns that the array does not distinguish signals arriving from the defined boresight front or from the rear. Increasing the number of elements in the array beyond three would improve the performance of the device for direction-finding. Figure 8.23 is a plot of the 45° pattern on a five-element array that shows a reduced sidelobe compared to the pattern in Figure 8.22(d). Full 360° coverage can be achieved by spacing antenna elements on a circle instead of in a line. With many elements and more sophisticated phase adjustment algorithms, sidelobes can be reduced and nulls tuned to cancel interfering signals [9].

The phase shift networks shown in Figure 8.21 can be conveniently implemented by creating a quadrature signal at each antenna element output, in addition

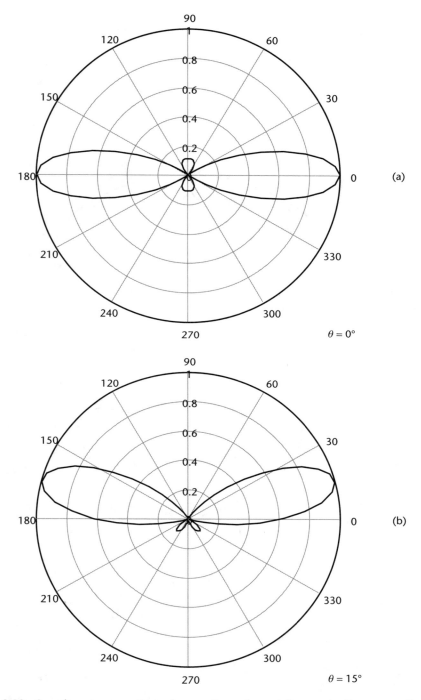

Figure 8.22 Sample antenna patterns from a three-element linear adaptive array adjusted for $\theta = 0°$, 15°, −30°, 45°. The radial scale is normalized linear power. (a) $\theta = 0°$, (b) $\theta = 15°$, (c) $\theta = -30°$, and (d) $\theta = 45°$.

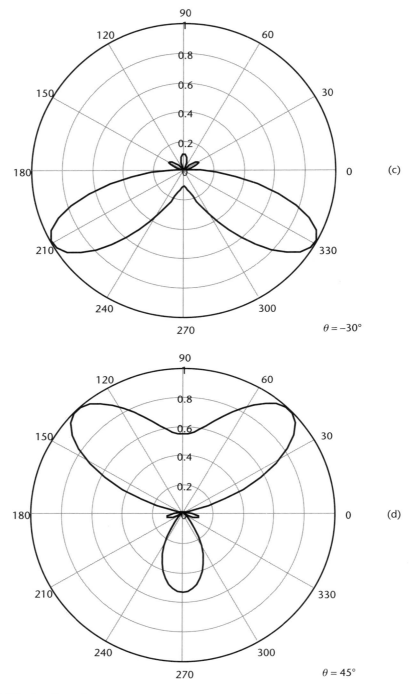

Figure 8.22   (continued).

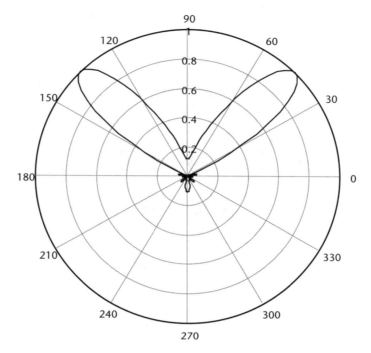

Figure 8.23 Antenna power pattern of a linear array with five elements and phase shift networks set for 45° angle of arrival.

to the direct signal [9]. The principle is illustrated in Figure 8.24. Each element has a 90° delay line or phase shifter, and the two orthogonal element outputs are tuned by a pair of adjustable amplitude weights. The sign and multiplying value of each weight is determined by the adaptive controller. The sum of the signals from each weight pair can have any phase shift between 0° and 360° relative to the phase of the direct signal from the antenna element, and a range of amplitudes. This can be seen by representing the weighting of an ith array element output as:

$$W_i = A_i \cdot \exp(j\alpha) = w_{i,2} + j \cdot w_{i,1} \tag{8.24}$$

where

$$w_{i,2} = A_i \cdot \cos(\alpha) \tag{8.25}$$
$$w_{i,1} = A_i \cdot \sin(\alpha)$$

and

$$A_i = \sqrt{w_{i,2}^2 + w_{i,1}^2} \tag{8.26}$$

The adaptive controller compares the output of the summing network to a desired criterion, such as maximum power, maximum signal-to-noise ratio, or lowest interference, and adjusts the weights in an iterative process until the criterion

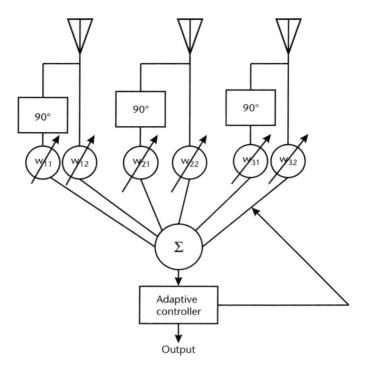

Figure 8.24 Phase adjustment using weighting blocks in an adaptive linear combiner.

is met. Usually, the adaptive process is aimed to attain a minimum mean square error between the summing network output and the desired result. After each iteration of adjusting the weights and measuring the mean square error, a process is performed to determine how to change the weights in the next cycle in order to reduce the mean square error and more closely approach the optimum values of the weights. There are several known algorithms for performing this cut and try approach systematically. Three of them are Newton's method, the method of steepest descent, and the LMS (least mean square) algorithm. They are described in detail in [9].

8.6   ESPAR Antenna Array

An adaptive direction finding antenna array that uses a different method of adjusting antenna element phase outputs is the electrically steerable parasitic array radiator (ESPAR) antenna [10, 11]. Instead of adjusting the weights of the outputs of several array members, ESPAR determines the antenna pattern by varying loading reactances in parasitic elements. A principal advantage of this method is that only one antenna element, a driven element, is connected to the receiver or transmitter. In the previous section, the antenna elements were assumed independent, that is, the phase of the currents in any element is not affected by the proximity of other elements. In the ESPAR antenna, the phase of the current in each element is dependent on the currents in the other elements of the array, and on variable loading reactances. The principle of closely placed antenna elements determining the antenna pattern is implemented in the Yagi antenna described in Section 8.3.

An ESPAR antenna is shown in Figure 8.25 [10]. Seven monopole elements are mounted on a ground plane. A center monopole is the driven element that is surrounded by six equally spaced elements located on a circle. Approximate dimensions are: elements $\lambda/4$ high and the radius of the circle also $\lambda/4$. The ground plane diameter is one wavelength and has a $\lambda/4$ conductive skirt width. The driven element is connected to the receiver. An adaptive controller adjusts the reactances of the parasitic elements either in an open loop to give a previously determined pattern, or in closed loop while using the RSSI output of the receiver for feedback. The variable element load reactances are generally controlled by voltage variable capacitors—varicaps. The reactance actually seen by an element can be mapped from the capacitance range of the varicap to values needed to form the antenna patterns by using transmission line impedance transformation techniques.

The equivalent circuit of the ESPAR array when used for receiving is shown in Figure 8.26. The voltage sources shown at the top of the diagram are the received signals at the individual elements, with relative phase of each a function of the direction of arrival from the target and the position of the element in respect to the wave front. The geometry of the relative phase difference between any two elements is the same as that shown in Figure 8.13. The admittance network Y can be represented by a matrix containing the self-admittances of each element and the mutual admittances between each element and all other elements. Y is a constant—a function of the physical makeup of the antenna including the ground plane. It depends principally on the length of the elements, their diameter, and the separation between them. The currents in all of the elements are determined by Y and by the impedance components connected to those elements. Element currents, except in the driven element, are adjusted by variable reactances. The current in the driven element is affected by the receiver input impedance to which it is connected.

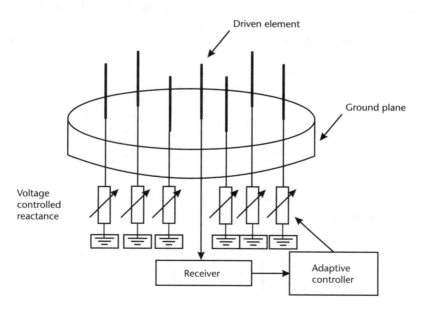

Figure 8.25   ESPAR antenna.

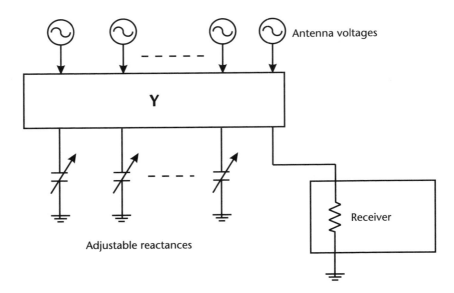

Figure 8.26 Equivalent circuit of ESPAR array in receive mode.

We have seen how adaptive antennas having separate inputs for each element are controlled by manipulating weighting factors on each signal, and then summing all signals to produce an output. It is desirable to map the reactance adjustment control of the ESPAR antenna to an equivalent weighting factor arrangement, so that beam steering algorithms for adjusting the antenna pattern can be expressed in a manner similar to that used with arrays having outputs from each element. Such a mapping can be carried out as follows.

The relationship between driving voltages, antenna currents, the antenna electrical admittance description Y, and the adjustable impedances is most conveniently developed for a transmitting antenna [10]. Figure 8.27 shows the configuration for N elements. Using matrix notation, the currents into the antenna elements are expressed as:

$$\mathbf{I} = \begin{bmatrix} i_0 \\ i_1 \\ \vdots \\ i_{N-1} \end{bmatrix} = \mathbf{Y} \cdot \begin{bmatrix} v_0 \\ v_1 \\ \vdots \\ v_{N-1} \end{bmatrix} \qquad (8.27)$$

The matrix $\mathbf{Y}$, whose members are the self and mutual admittances between the N elements, is

$$\mathbf{Y} = \begin{bmatrix} y_{0,0} & y_{0,1} & \cdot & y_{0,N-1} \\ y_{1,0} & y_{1,1} & \cdot & \cdot \\ \cdot & \cdot & \cdot & \cdot \\ y_{N-1,0} & y_{N-1,1} & \cdot & y_{N-1,N-1} \end{bmatrix} \qquad (8.28)$$

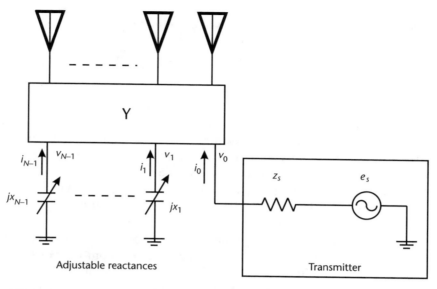

Figure 8.27 ESPAR antenna in transmitting mode.

The voltages v_1 through v_{N-1} are the element currents times the adjustable reactances whereas $v_0 = e_s - i_0 z_0$. Those reactances and the source impedance of the transmitter are represented by a diagonal matrix $X = \text{diag}(z_s\ jx_1\ jx_2 \ldots jx_{N-1})$. We also need a convenience vector U of order N so that the source voltage can be included in the matrix equation (8.27):

$$U = [1 \quad 0 \quad \ldots \quad 0]^T \tag{8.29}$$

Now the voltage vector on the right side of (8.27) is replaced by a vector showing the element currents, adjustable reactances and the excitation (transmitter) voltage, giving:

$$I = Y \cdot (e_s \cdot U - X \cdot I) \tag{8.30}$$

Through simple manipulation using matrix algebra, the I vectors are collected:

$$I = e_s \cdot (Y^{-1} + X)^{-1} \cdot U \tag{8.31}$$

$$= \frac{e_s}{2z_s} \cdot 2z_s \cdot (Y^{-1} + X)^{-1} \cdot U$$

The second form of (8.31) expresses the antenna element currents I as multiples of a reference current, $e_s/(2z_s)$, which is the current into the array from the transmitter when the input impedance equals the source impedance. The multiplication factor is a unitless vector W:

$$W = 2z_s(Y^{-1} + X)^{-1} \cdot U = [w_0 \quad w_1 \quad \ldots \quad w_{N-1}]^T \tag{8.32}$$

Now we can rewrite (8.31) as a voltage equation:

$$\mathbf{E} = \mathbf{I} \cdot 2z_s = e_s \cdot \mathbf{W} \tag{8.33}$$

where $\mathbf{E} = [e_0 \quad e_1 \quad \ldots \quad e_{N-1}]^T$.

$\mathbf{W}$ defined in (8.32) is called the *equivalent weight vector*. Using (8.33), the system of Figure 8.27 can be represented as shown in Figure 8.28 where the relative phases of the antenna elements, and thus the antenna pattern, are determined by the equivalent weight vector. The radiating and receiving properties of an antenna are reciprocal, and a representation of the ESPAR antenna for receiving, corresponding to the transmitting mode of Figure 8.28, is drawn in Figure 8.29. Figure 8.29 is very similar to Figure 8.24. The weights in (8.33) and Figure 8.28 are complex, and their real and complex parts could be shown exactly as in Figure 8.24, with 90° phase shifts from the individual antenna elements. The major difference between the ESPAR representation and the adaptive linear combiner of Figure 8.24 is that the weights of the latter are controlled directly by an adaptive controller whereas in Figure 8.28 and Figure 8.29 the weights are an abstract concept and actual

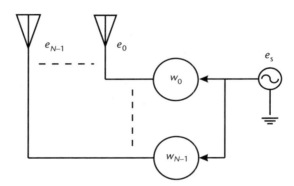

Figure 8.28   Representation of ESPAR antenna in transmission mode with element voltages controlled by equivalent weights.

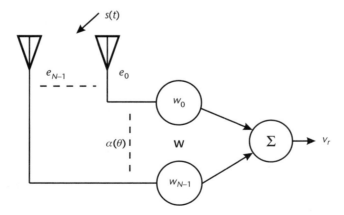

Figure 8.29   Representation of ESPAR antenna in receive mode with pattern controlled by equivalent weights.

control is through adjustable reactance elements. Also, the antenna in Figure 8.24 has separate RF inputs and amplifying channels from each element at the end of which is the summing device. The ESPAR antenna element signal summation is inherent to the array and there is only one RF connection to the receiver or transmitter.

The antenna pattern of the ESPAR receiving antenna is adapted conceptually by adjusting the weight vector that is applied to the antenna element voltages to get the desired results. The phases of the voltages induced on the antenna elements from the received signal depend on the direction of arrival of the signal and the physical arrangement of the antenna elements. The steering vector, $\boldsymbol{\alpha}(\theta)$, expresses the relationship between the antenna voltage phases and the direction of arrival. Figure 8.30 displays a plane layout of the elements of the antenna pictured in Figure 8.25, viewed from above. The numbered elements are mounted symmetrically on a circle of radius $\lambda/4$ around the driven element 1. The reference axis for the angle of arrival is through elements 1 and 2. When a signal arrives from a target on an angle of θ, the phase of the voltage induced in element 1 lags the phase of the voltage in element 2 by $2 \cdot \pi \cdot \dfrac{a}{\lambda} = \dfrac{\pi}{2} \cdot \cos(\theta)$. The phases of the other elements in respect to the phase in element 1 are found similarly and so the steering vector, expressed in complex form, is

$$\boldsymbol{\alpha}(\theta) = \begin{bmatrix} 1 \\ \exp\left(j \cdot \dfrac{\pi}{2} \cdot \cos(\theta)\right) \\ \exp\left(j \cdot \dfrac{\pi}{2} \cdot \cos\left(\theta - \dfrac{\pi}{3}\right)\right) \\ \exp\left(j \cdot \dfrac{\pi}{2} \cdot \cos\left(\theta - \dfrac{2\pi}{3}\right)\right) \\ \exp\left(j \cdot \dfrac{\pi}{2} \cdot \cos(\theta - \pi)\right) \\ \exp\left(j \cdot \dfrac{\pi}{2} \cdot \cos\left(\theta - \dfrac{4\pi}{3}\right)\right) \\ \exp\left(j \cdot \dfrac{\pi}{2} \cdot \cos\left(\theta - \dfrac{5\pi}{3}\right)\right) \end{bmatrix} \tag{8.34}$$

The steering vector for the antenna determines the voltages on the elements for an incoming signal $s(t)$. The signal at the receiver RF input (Figure 8.29) is

$$v_r(t) = \mathbf{W}^{\mathrm{T}} \boldsymbol{\alpha}(\boldsymbol{\theta}) s(t) \tag{8.35}$$

If the weights $\mathbf{W}$ are adjusted so that the phases of the signals at the input to the summing block of Figure 8.29 are all equal, v_r will be maximum and the value of $\mathbf{W}$ can be interpreted to deduce the angle of arrival. As an alternative that will

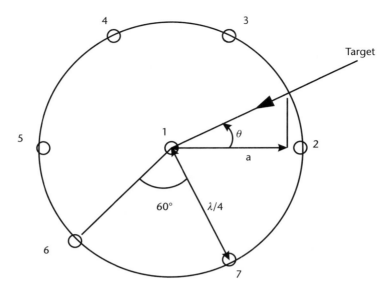

Figure 8.30 Layout of seven-element ESPAR antenna.

give a sharper distinction of the angle, $\mathbf{W}$ can be adjusted for a null at v_r, which, too, can be translated into angle of arrival.

As mentioned above, in the ESPAR antenna there is no access directly to the weights $\mathbf{W}$—they must be mapped from the reactance values as shown in (8.32). The physical entity that is actually varied by the adaptive controller of the ESPAR antenna is the voltage on the varicaps, which form the adjustable reactances. While the concept of the abstract weight vector is very useful for understanding how the ESPAR antenna works, the algorithm for adjusting the antenna pattern doesn't use $\mathbf{W}$ directly. Beamforming control works by an iterative process during which the voltages on the varicaps are changed in steps until the best possible result at v_r is obtained. After each step, a performance criterion is examined and the algorithm decides how to change the voltages for the next step [10]. The range of reactances that can be covered by the varicaps, as well as the degree of influence of the element reactances on the antenna pattern, are limited and the coverage of the equivalent weights of ESPAR is less than can be achieved for an adaptive antenna whose elements are connected to separate receiving channels. However, by using a minimum least mean square error search routine in software that takes into account the limitations of the antenna pattern adjustment, useful results have been obtained [10].

8.7 Conclusions

The angle-of-arrival method of wireless location has several advantages compared to time-of-flight methods. It does not require cooperation with the target emitter nor is it dependent on particular modulation characteristics or stringent receiver timing. For its simplest implementation it relies on highly directional antennas with narrow beam widths, as well as a rotation mechanism. Such antennas have physical dimensions of several wavelengths and may be practical only for microwave bands. As an alternative to mechanical rotation, direction finding can be accomplished using amplitude or phase comparison techniques.

The ESPAR antenna is a relatively compact array that achieves beam scanning by electrically changing the reactances of parasitic array elements. Generally the range of pattern control is limited as compared to that possible with arrays having elements with feed points directly accessible to the receiver. The advantage of ESPAR is that only a single receiver or transmitter RF channel is required and pattern adjustment is done through voltage control of reactances on the parasitic elements.

Noise and interference limit location accuracy of AOA systems, as with systems using alternate positioning methods. The classical AOA technique of highly directional arrays has the advantage of inherently rejecting interference from directions away from the target, as well as boosting signal-to-noise ratio. The amplitude comparison and phase interferometer techniques suffer from directional ambiguities and increased susceptibility to interference. Sophisticated processing techniques increase location resolution and reject interference and multipath distortion. Phase comparison techniques are basically suited to narrowband signals and special processing algorithms are needed for adapting them to wideband channels.

References

[1]   Roy, S., et al., "Neighborhood Tracking and Location Estimation of Nodes in Ad Hoc Networks Using Directional Antenna: A Testbed Implementation," *WirelessCom 2005*, Maui, HI, June 13–16, 2005.

[2]   Hansen, R. C., "Antennas," in *Reference Data for Engineers*, 9th ed., W. M. Middleton, (ed.), Boston, MA: Newnes, 2002, p. 32–11.

[3]   Landee, R. W., D. C. Davis, and A. P. Albrecht, *Electronic Designers' Handbook*, New York: McGraw-Hill, 1957.

[4]   Hall, G., (ed.), *The ARRL Antenna Book*, Newington, CT: The American Radio Relay League, 1991, p. 18-8.

[5]   Terman, F. E., *Electronic and Radio Engineering*, New York: McGraw-Hill, 1955, p. 914.

[6]   "AD8302 RF/IF Gain and Phase Detector Specification," Analog Devices, Inc., Norwood, MA, 2002.

[7]   Yoon, Y. S., L. M. Kaplan, and J. H. McClellan, "DOA Estimation of Wideband Signals," in *Advances in Direction of Arrival Estimation*, S. Chandran, (ed.), Norwood, MA: Artech House, 2006.

[8]   Abe, J., et al., "Real-Time Location Estimation System for Wild Animals," http://www.ap. ide.titech.ac.jp/publications/Archive/IEEJ_IM(0606Abe).pdf, 2006.

[9]   Widrow, B., and S. D. Stearns, *Adaptive Signal Processing*, Englewood Cliffs, NJ: Prentice-Hall, 1985.

[10]   Sun, C., et al., "Fast Beamforming of Electronically Steerable Parasitic Array Radiator Antennas: Theory and Experiment," *IEEE Transactions on Antennas and Propagation*, Vol. 52, No. 7, 2004.

[11]   Taromaru, M., and T. Ohira, "Electronically Steerable Parasitic Array Radiator Antenna— Principle, Control Theory and Its Applications," *28th General Assembly of International Union of Radio Science (URSI GA 2005)* New Delhi, India, October 23–29, 2005.

CHAPTER 9

Cellular Networks

Cellular networks are the major platform for wireless location-based services (LBS). This is natural considering the widespread distribution of cellular handsets among a large part of the world's population. Position accuracy demands and the methods used for positioning are highly dependent on the nature of these services. Most of the methods already discussed in detail in previous chapters have been applied to technologies for adding location capability to cellular systems. They include TOF (in the form of TDOA), AOA, and RSS. Multifrequency techniques have not been used because they are not generally adaptable to the air interfaces up to third generation technologies. The adoption of OFDM for cellular systems in future evolution of technologies may very well add multifrequency techniques to the repertoire of methods that are already being used for the various network types and applications.

9.1 Cellular Location-Based Services

The driving reason for adding location capability to cellular communication was physical security for the holders of handsets, at least in the United States where cellular providers are obligated by telecommunication regulations to provide positioning as a nonsubscription service. Once the infrastructure and/or handset models were available for providing location, it was only natural the range of services based on location would start to bloom. In Europe and other world regions it is these commercial services that are generating the inclusion of location capability in cellular networks. Some of the most common location-based services, other than personal security, are the following:

- Provision of transport navigation instructions;
- Identification of nearby commercial institutions—restaurants, banks, hotels, and so forth—as an adjunct to navigation;
- Tracking of people, animals and things—child finder and stolen vehicle recovery, for example;
- Location sensitive cellular billing;
- Fraud detection in use of cellular network;
- Cellular system design and resource management, and improved performance;
- Fleet management and intelligent transport systems (ITS).

In 1996, the U.S. Federal Communications Commission (FCC) issued a Report and Order requiring that specified commercial mobile radio service providers, including cellular, provide information to emergency 911 (E-911) public safety services. (We include here under "cellular" both systems operating on the 800-MHz bands and the personal communications services (PCS) operating on 1,800–1,900-MHz bands.) This was an extension to the service available to all landline telephone subscribers that calls made by dialing 911 to a Public Safety Answering Point (PSAP) would be automatically accompanied by the location of the calling party. Obviously, determining the location of a mobile cellular handset without involving the caller is much more involved than getting the same information from a fixed phone user. Therefore, introduction of the mandatory location identification service was to be done gradually, starting with provision of the location of the individual network base station cell that is in contact with the caller. At a later date, the service provider was to determine the position of the caller within a defined location accuracy. The realization of the FCC Order was apparently more difficult, technically, than was assessed by the FCC and the providers, and the schedule was delayed, while the accuracy requirements were also modified.

The FCC requirements regarding the provision of location for wireless 911 dialed calls are defined in Chapter 47, Part 20, Section 20.18 of its regulations [1]. They include two phases. Under the Phase I enhanced 911 services the supplier of wireless communications must provide the PSAP with the telephone number of the 911 caller and the location of the cell site or base station that receives the call from a mobile handset. This requirement is quite simple to implement as it doesn't involve any wireless distance measuring methods or information beyond that which it has in any case: the location of the cell with which the handset is communicating. The Phase II requirement, on the other hand, can be complied with only by using a distance measuring method or technology that estimates to a specified degree of accuracy the location of the handset making the 911 call. Those providers subject to the Phase II service requirements must pass on to the PSAP the location of 911 calls in terms of latitude and longitude in conformance with the following accuracy requirements. For network-based technologies (defined later) the accuracy is within 100m for 67% of calls and 300m for 95% of calls. For handset-based technologies (defined below) the accuracy is 50m for 67% of calls and 150m for 95% of calls. For the remaining 5% of calls, location attempts must be made and a location estimate for each call must be provided to the appropriate PSAP [1].

In Europe, emergency calls are made by dialing 112 in most countries and the services provided are called Enhanced 112 services [2]. The cellular network provides caller location details to emergency authorities, but there are no specifications for accuracy or distinction between mobile and fixed callers. Generally, location technologies are designed to meet requirements for specific location-based applications, rather than to meet mandatory specifications as in the United States

9.2 Cellular Network Fundamentals

In order to understand how cellular radio positioning works, some understanding of cellular network fundamentals is required. The following discussion concentrates on the aspects of cellular networks that apply to the provision of location services. The examples are taken from the pan-European cellular system, GSM, and the North American CDMA IS-95 standard [3]. General network operation principles are similar for all cellular systems although the air interfaces including operating frequency bands differ.

A simplified block diagram of a cellular phone network is shown in Figure 9.1. The terminology is that of Global System for Mobile communication (GSM). Mobile stations (MS) communicate directly with base transceiver stations (BTS). The base station subsystem (BS) includes the BTS and a base station controller (BSC) that governs the air interface parameters of a cell or cell sector, including frequency and power control, broadcast traffic control, and handover initiation. Networks that have a TDOA location service also include in the base station subsystem a location measurement unit (LMU) that collects time of arrival data from an MS target. The mobile switch center (MSC) serves as the gateway to the fixed public network that includes the public switched telephone network (PSTN) as well as the integrated services digital network (ISDN) and packet data network (PDN). It also provides access to location registers where data on the mobile stations in the network are stored, and to the authentication function.

At the block diagram level, all cellular systems are quite similar, but their air interfaces differ significantly, and it is the air interface that predominantly affects the performance of the location function. A comparison of several parameters of the air interfaces of second generation GSM and CDMA IS-95, and third generation WCDMA (UMTS) is given in Table 9.1.

The transmission direction between mobile stations and base stations is referred to in two ways. When the base station is considered the origin, or reference point, a forward channel is a communication link on which data flows from the BS to the MS. On a reverse channel, data flows from MS to BS. Considering the MS as the reference point, the downlink direction is from BS to MS, and uplink data

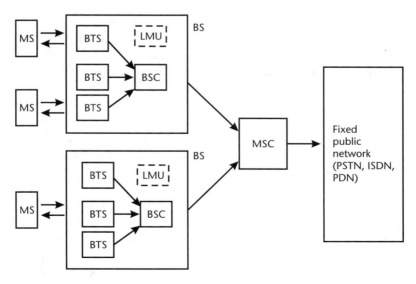

Figure 9.1   Cellular phone network.

Table 9.1 A Comparison of GSM, CDMA IS-95, and WCDMA (UMTS) Air Interfaces

Feature	GSM		CDMA IS-95		WCDMA (UMTS)	
Major frequency bands (MHz)	Uplink 890–915 1,710–1,785 1,850–1,910	Downlink 935–960 1,805–1,880 1,930–1,990	Uplink 824–849	Downlink 869–894	Uplink 1,920–1,980	Downlink 2,110–2,170
Symbol/chip rate (kbps)	270.8		1,288		3,840	
Bit period (μs)	3.69		0.776		0.260	
Channel width (kHz)	200		1,250		5,000	
Multiple access	Time division (TDMA)		Code division (CDMA)		Code division (CDMA)	
Modulation	GMSK (Gaussian minimum shift keying)		Phase shift keying		Phase shift keying	
Power control	Yes		Yes		Yes	

flows from MS to BS. A handset-based location system operates on downlink data whereas a network-based system measures characteristics of the uplink signal.

Data between the MS and BS is arranged in a hierarchy of frames and time slots. Communication is carried out over physical channels that are classified into traffic channels and control channels. Traffic channels contain the information, speech, or data that is transferred between a mobile terminal in the network with a terminal in any other fixed or cellular network after a call is set up. Control channels are maintained between mobile and base station in order to set up and terminate calls, to synchronize slot time and frequency assignments, and to facilitate handover between mobile and adjacent cells.

9.2.1 GSM Transmissions

GSM uses time division multiple access (TDMA) and transmissions are conducted in bursts of duration 577 μs, including a silent guard period. Each burst fills a time slot, and there are eight time slots in a TDMA frame. A mobile station requesting access to the network is assigned one of the eight time slots that has not been assigned to another terminal. Thus, up to eight MSs can be registered simultaneously on the same frequency. Due to the propagation time between mobile and base station, it would be possible for transmissions from mobile stations to overlap when received at the base station. In the same manner, the propagation delay could cause the mobile to hear a BS transmission that is intended for a different MS. The effect of propagation delay on slot alignment is shown in Figure 9.2. The mobile station sets its clock and time base counters to those of the strongest base station in the vicinity by reception of a synchronization burst from the base station. As Figure 9.2 shows, slot alignment is skewed from that of the BS by the propagation time τ between the stations. The MS replies to the BS, which notes the number of bits of delay from the start of the GSM frame. This bit delay is then twice the propagation delay. The BS sends the MS this bit delay as a timing advance (TA), which the MS uses to adjust its slot timing so that its transmissions are correctly received by the BS and so that it receives time slots intended for it from

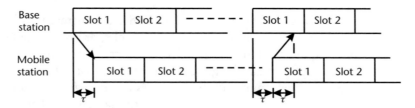

Figure 9.2 GSM time slot skew due to propagation delay.

the BS. Timing advance has a resolution of 1 bit, 3.69 μs, and its range is zero to $63 \times 3.69 \ \mu s = 232.5 \ \mu s$, corresponding to a maximum two-way distance of around 70 km, or a range of 35 km. The timing advance then is the basic distance measuring feature of GSM, which can be use to estimate distance between BS and MS to a resolution of $0.5 \times 3.69 \ \mu s \times 3 \times 10^8$ m/s = 554m.

Cellular TDOA location technologies use transmission burst features to estimate time of arrival. Two of the four types of bursts in GSM are used for this purpose. A normal burst, used for traffic channel and most control channel transmissions, is shown in Figure 9.3 [3]. The 26-bit training sequence in the middle of a received burst is cross correlated by the known sequence in the receiver. A handset-based technology can likewise use the 64 training sequence, Figure 9.4, in a synchronization burst from a base station to make its TOA estimation. Similar measurements made in handsets from several neighboring base stations or by several LMUs from a handset transmission can be used to find location by the TDOA method of estimating the intersection of hyperbolas. Oversampling is needed to improve the one bit resolution of the TA to obtain a precision of tens of meters that is required to comply with FCC E-911 requirements.

9.2.2 CDMA

Code division multiple access (CDMA) cellular systems are based on direct sequence spread spectrum principles and ways of obtaining times of arrivals of epochs of signals are similar to those discussed in Chapter 3. However, in order for their multiple access scheme to be effective, strict power control is used to overcome

Tail bits	Encrypted data	Training sequence	Encrypted data	Tail bits	Guard time
3	58	26	58	3	8.25

Figure 9.3 GSM normal burst. (*After:* [3].)

Tail bits	Encrypted data	Training sequence	Encrypted data	Tail bits	Guard time
3	39	64	39	3	8.25

Figure 9.4 GSM synchronization burst. (*After:* [3].)

what is called the near-far field effect [4]. In CDMA, transmissions of all subscribers in a single cell, and often those of neighboring cells, are conducted simultaneously on the same frequency channel. The spread spectrum principle of processing gain allows reception of a desired signal in the presence of interference on the same frequency, but the maximum number of users is contingent upon the necessity for all received signals at the base station to have the same level. Thus, mobile stations located close to a BS must reduce their power, and others in the same cell that are located far from the BS must set their power close to maximum. TDOA is based on the necessity, in network-based location, to estimate TOA for a particular MS at several fixed, geographically separated, transceivers. If an MS is located close to the BS of the cell it is in, its power will be reduced, and therefore may not be heard with a sufficient signal-to-noise ratio at location service receivers in other cells, whose TOA data from the MS is required to calculate the target position. The near-far problem is alleviated during handover between base stations. During handover, the power control is temporarily stopped, and there is an opportunity for enough location measurement units to make the TOA measurements needed for the calculations.

In handset-based location systems, the pilot tone signal transmitted at constant power from all base stations can be used to obtain TOA data. The handset must have special software to cause it to measure TOA from the base stations in its vicinity. It can then send these measured data to a fixed station that knows the location coordinates of the base stations and can calculate the position of the MS from the intersection of hyperbolas.

9.2.3 UMTS

Location accuracy from third generation cellular is significantly better than that achieved in GSM and CDMA. Inherent accuracy is higher because of the increased signal bandwidth and shorter bit period (Table 9.1). These features significantly improve the ability to distinguish the line-of-sight signal among multipath returns in reception. Universal Mobile Telecommunication System (UMTS) supports the following location methods [5]:

- Cell-ID;
- Observed TDOA-idle period downlink (OTDOA-IPDL);
- Network-assisted GPS (A-GPS);
- Uplink TDOA (U-TDOA).

In UMTS, timing advance is referred to as round-trip time (RTT). RTT is reported with a resolution of 1/16 chip, or approximately 5m. Forced handovers are used to eliminate the near-far problem caused by power control. Figure 9.5 shows UMTS architecture applicable to the U-TDOA location method. Requests for location of the user equipment (UE) from the core network (CN) are managed by the radio network controller (RNC) according to UMTS protocol. TOA data is taken by the location measurement units and processed in the stand-alone serving mobile location center (SAS). The LMUs are connected to the SAS over an overlay network. The RNC can control the uplink and downlink powers as required for the position-

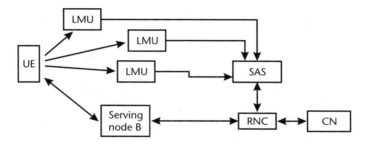

Figure 9.5   UMTS positioning architecture for U-TDOA.

ing function in order that, in the case of U-TDOA, EU transmissions are heard by the required number of base station transceivers (node B or LMU). The function of the LMU is to create the TOA data from the mobile signal in the network based method and time stamp it so that time difference values can be found with the data from other LMUs participating in the location task. Time measurements are related to a common clock, such as provided by GPS receivers at each LMU location, or can be relative to a particular node B [5]. Calculations and determination of position output are done in a serving mobile location center (SMLC). The SMLC manages the resources required for mobile unit location and controls a number of LMUs for the purpose of obtaining radio interface measurements. The SMLC may be stand alone (SAS) or part of a RNC. The actual measurement flow and commands depend on the method of location that is being used, which may be A-GPS, OTDOA, Cell-ID, or U-TDOA.

9.3   Categories of Location Systems

The two major categories for describing location systems are network-based and handset-based. Network-based systems use technologies that determine handset position solely from measurements taken at cellular base stations, with no requirements from the handset other than those necessary for the normal communication function. Such handsets are called legacy handsets. A network-based location system estimates the location of any handset that operates in the network. In a handset-based location system, location measurements are made in the handset and special software, and often hardware, must be incorporated in the handset to give it the ability to estimate its location. Also defined are hybrid systems where both the handset and the network are modified to accommodate the positioning function.

Both handset-based systems and network-based systems have their advantages and disadvantages in relation to particular applications. Following are the most important advantages of each category.

Handset-Based

- When the position information is used by the target itself, handset-based location is most secure. Location information and tracking of the target are not available in the network.

- Network capabilities are not involved, and a roaming handset can be used in any communications compatible network.
- The system does not use facilities and resources of the network, so network capacity is not affected.
- The handset is not limited by the network on the number of measurements it can take, so location accuracy can be improved as required by taking more measurements.

Network-Based

- All legacy handsets can receive location services without subscribers having to upgrade their device.
- The network has more computing power than the handset so it can take advantage of positioning methods that would be impractical at a handset.
- A network-based system frees the handset of the battery power penalty of having to carry out positioning tasks.
- The system can initiate target positioning and tracking without intervention or action by the target.

A third category is a hybrid positioning system where both the handset and the network are involved in the positioning function. An example is the case where measurements are made by the handset and transmitted to the network that performs the location calculations.

There is another way to classify location systems [6]. *Self-positioning* refers to a system where the target itself independently takes measurements and calculates its own location. The best example of a self-positioning system is a handset with a GPS receiver. In a *remote positioning system*, target device transmissions are used by multiple receivers to find the target's location. Self-positioning is generally synonymous with handset-based and remote positioning with network-based.

9.4 GPS Solution

A common handset-based self-positioning system is a handset that includes a GPS receiver. The GPS portion of the handset is independent of the network and provides the coordinates of the handset just like any other self-contained GPS receiver. Those coordinates can be transmitted over the cellular network as SMS or packet data (GPRS) and used for location-based services. Such a solution has the possible advantage of being completely independent for its operation from the cellular network, but is accompanied by some negative aspects. It needs all the computing power of a stand-alone GPS radio, which may be expensive to support in a cell phone handset, and will also cause increased current drain from the handset battery. Due to the nature of use of cell phones, it may not have access to all satellites that are available at the handset location because of blocking, particularly in urban and indoor environments. Since the GPS function will not be operating all of the time in order to maintain handset battery charge time, time to first fix on actuating the location function may be inordinately long for many LBS applications.

These problems are for a large part alleviated by an assisted GPS (AGPS) solution. In AGPS, many of the functions of a full GPS receiver are performed by a remote GPS location server. These are the characteristics of AGPS. The remote server provides the AGPS handset with:

- Precise satellite orbit and clock information;
- Initial position and time estimate;
- Satellite selection, range, and range rate;
- Position computation.

As a result, the handset contains a very basic GPS receiver that needs only to synchronize to given satellites that are visible to it, and then transfer pseudotime difference or range to the location server over the cellular network. It is not required to decode the GPS messages for each satellite or to perform an extensive search for visible satellites when the system is turned on. Synchronization time is reduced and sensitivity can be significantly increased thereby enabling use of partially blocked satellites whose signal could not be used if full message decoding was required. Thus, AGPS handsets have reduced power consumption, rapid location determination, and the ability in many cases to function indoors where a full GPS handset receiver would be unable to obtain a fix. AGPS has been adopted by cellular service providers to satisfy the FCC requirements for provision of E-911 service. The process has the following steps:

- Upon initiation of a 911 call, the cellular network sends the handset the approximate location in the form of the serving cell identification.
- The location server tells the handset GPS function what satellites are in view at that location.
- The handset GPS receiver synchronizes with the known satellites in view and sends pseudorange data to the location server.
- The location server performs error corrections and calculates the handset position coordinates and sends them in specified format to the PSAP. For other applications, the server can send the position data back to the handset or to a third-party location-based service provider.

9.5   Cell-ID

The most basic positioning technology available for cellular systems is called cell-ID (cell identification). It can be either handset-based or network-based. In order to conduct communication, a handset is associated with an individual base transceiver located in a network cell. The cell identity and location are of course known to the base station. The cell identity is also known to the handset, which can obtain the cell's position from the network. This location method is referred to as *proximity*. Its basic accuracy depends solely on the dimensions of the cell, but can be enhanced by support of other location methods. A reduction of the location area and therefore increased resolution is achieved in cells that are divided into three or six sectors by directional base station antennas. This is shown in Figure 9.6(a). With 120° beam width antennas, the cell area and therefore the inaccuracy of the cell-ID technology is reduced by approximately one-third. Location accuracy can be improved by using the received signal strength as in the example of Figure 9.6(b). Signal strength varies considerably, as mentioned in Chapter 6, due to fading, topography, antenna patterns, radiated power, and operating frequency. At least the power and frequency factors are stabilized when the handset reads the signal strength of the broadcast channel from the base station.

Another way to limit the location uncertainty in the cell-ID technology is to use the timing advance that is calculated by the base station in GSM systems. The timing advance is the correction that the network makes to the mobile station timing to account for propagation delay so that the slots allocated to MSs do not overlap. The resolution of the timing advance in GSM is 3.69 μs, equivalent to a one-way distance of 554m. Figure 9.6(c) shows how position uncertainty is reduced when timing advance is combined with cell-ID to estimate location.

9.6 Location Technologies Using TDOA

Several location technologies have been developed that are based on the TDOA method of distance measurement. Their operational details depend on whether

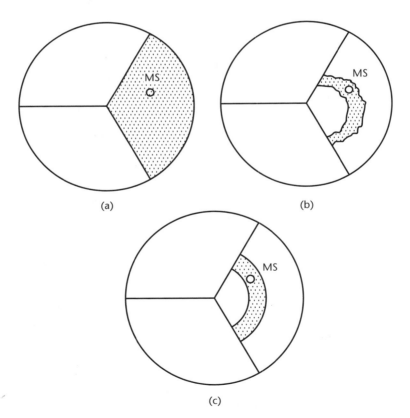

(a)   (b)

(c)

Figure 9.6 Cell-ID method with enhancements: (a) sectored cell, (b) cell-ID with RSS, and (c) cell-ID with timing advance.

they are handset-based or network-based, and how they are applied to a specific network protocol. In all cases they depend on estimation of time of arrival of an epoch in the cellular signal [4]. Time of arrival is found by finding the maximum cross correlation of a received signal with a delayed replica of the known transmitted signal. Within the transmission frames of cellular signals are synchronization or training sequences that are known to the receiver, and that have low autocorrelation when the signal and its replica are not lined up bit for bit. Let $s_1(t)$ be the transmitted sequence. The received sequence is $s_2 = s(t - \tau) + n(t)$ where $n(t)$ is noise and τ is the propagation delay. The cross correlation is

$$R(t') = \frac{1}{T} \int_0^T s_1(t) s_2(t + t') \, dt \qquad (9.1)$$

The integration is performed on sampled sequences s_1 and s_2. The value of t' for which $R(t')$ is maximum is an estimate of the delay of s_2 in respect to s_1. The clocks of the handset and base station are not synchronized so only relative time delays from different stations can be estimated. In a handset-based system, the handset estimates the relative time delays t_i' of input sequences from three or more base stations, as compared to the time of the known sequence that is based on the handset clock. Often there is no synchronization between the clocks of different base stations. In order to estimate its own position using the TDOA method of intersecting hyperbolas (in two dimensions) the handset must receive from an external source the transmission times of the sequences from each of the base stations according to a common clock, or the differences between them, and also the geographical positions of those base stations. As an alternative, the handset may send its observed time differences $\Delta t_{ij} = t_i' - t_j'$ to a special fixed terminal that has the timing and base station position information and can use those observed time differences to calculate the handset's position.

A network-based positioning system can also use TDOA to estimate the location of the mobile station. The sequence for cross correlation that is transmitted by the MS must be received by at least three base stations. The received times of arrival that are calculated by the base stations using (9.1) are then sent to a location estimating function which knows the positions of the receiving base stations and can than estimate the location of the mobile.

Perhaps the biggest problem in implementing a TDOA cellular location system is assuring that multiple base station receivers can hear the mobile in a multilateral network-based system, or in a unilateral system that the mobile station hears multiple base station transmissions. The unilateral case is generally easier to achieve, since the base stations send broadcast control signals that are used by the mobile station to determine what cell it is in. In a network-based system, the protocol must be adapted specifically for the location function so that the power control of mobile transmissions is temporarily aborted or otherwise made to comply with the necessity to be heard by multiple base stations.

The concept of a handset-based TDOA location system is shown in Figure 9.7. The MS receives sequences for correlation from each of the three base stations at different times. The LMU receives these same transmissions. Knowing its own

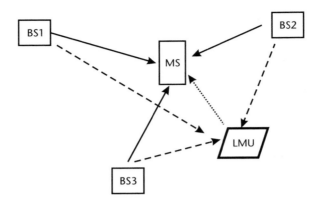

Figure 9.7 Layout of handset-based TDOA system.

position and the positions of the base stations, the LMU calculates the transmission times of the sequences from the base stations relative to its own clock. Let these times be t_{T1}, t_{T2}, and t_{T3}, for *BS1*, *BS2*, and *BS3*. The MS receives the sequences, using (9.1), at times t_{R1}, t_{R2}, and t_{R3}. The three time differences of arrival are

$$\Delta_1 = (t_{R2} - t_{R1}) - (t_{T2} - t_{T1})$$
$$\Delta_2 = (t_{R3} - t_{R2}) - (t_{T3} - t_{T2}) \tag{9.2}$$
$$\Delta_1 = (t_{R3} - t_{R1}) - (t_{T3} - t_{T1})$$

The location of the MS is found by solving for the best solution of the equations (see Chapter 7):

$$\sqrt{(y_2 - y)^2 + (x_2 - x)^2} - \sqrt{(y_1 - y)^2 + (x_1 - x)^2} = c \cdot \Delta_1$$
$$\sqrt{(y_3 - y)^2 + (x_3 - x)^2} - \sqrt{(y_2 - y)^2 + (x_2 - x)^2} = c \cdot \Delta_2 \tag{9.3}$$
$$\sqrt{(y_3 - y)^2 + (x_3 - x)^2} - \sqrt{(y_1 - y)^2 + (x_1 - x)^2} = c \cdot \Delta_3$$

where (x_1, y_1), (x_2, y_2) and (x_3, y_3) are the known coordinates of BS1, BS2, and BS3, and (x, y) are the coordinates of the MS that are to be estimated.

Several technologies have been developed that use the TDOA method for cellular mobile station location. They are described next.

9.6.1 Enhanced Observed Time Differences

Enhanced observed time differences (E-OTD) was an early technology applied to handset-based positioning in GSM networks. Signal flow is shown in Figure 9.8. As opposed to the operation of Figure 9.7, the handset does not calculate its own location. Observed time difference measurements from downlink signals from several base stations are routed from the MS to a mobile location center (MLC) that performs the calculation for estimating the MS position. GSM-based station

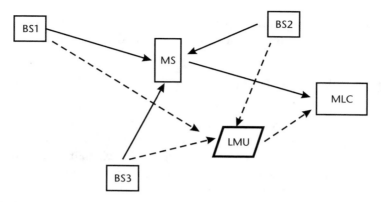

Figure 9.8 E-OTD architecture.

time bases are not synchronized, so accurate base station sequence transmission time differences are determined from measurements made by separate fixed receivers, often located at BS sites, one of which is indicated in Figure 9.8 as a location measurement unit (LMU). Handsets that are modified to perform E-OTD positioning have software algorithms for improving accuracy, by rejecting multipath responses, for example, However, the technology was not found to be sufficiently accurate for meeting E-911 requirements [2].

9.6.2 Observed Time Difference of Arrival

Observed time difference of arrival (OTDOA) is similar to E-OTD in that it is a handset-based technology using the TDOA method. It is used in third generation UMTS networks. In its *assisted mode*, position calculations are carried out in a serving radio network controller (SRNC) that is a mandatory part of the mobile terminal [7]. Time-of-arrival measurements at an UE (UMTS mobile station, Figure 9.5) taken from downlink messages from three or more node B terminals (UMTS base stations) are forwarded to the serving RNC. The relative time differences (RTD) of the transmission times of the downlink signals to the UE sent from the base stations are measured by LMUs and sent to the SRNC. The SRNC calculates the location of the UE using the OTDOA measurement data, the relative transmission time differences, and the coordinates of the node B terminals.

Also defined, in addition to the UE-assisted mode, is a *UE-based mode* in which the UE itself does the position calculations based on its TDOA observations from several base stations. It also must receive from the network the coordinates of the base stations from which it makes the TOA measurements, and the relative transmission time differences.

UMTS networks have a CDMA air interface on which neighboring base stations transmit concurrently on the same radio channel. Mobile stations (UEs) may have difficulty hearing the necessary number of base stations needed for the TDOA calculations, because the transmissions of the closest station, usually the one the UE is associated with, are stronger than those of the other stations participating in the OTDOA location estimation. One solution is an idle period downlink (IPDL) where the serving node B provides idle transmission periods to allow the mobile to receive other base stations that are located farther away [6].

9.6.3 Uplink Time Difference of Arrival

As compared to the two previously described TDOA handset based technologies, uplink time difference of arrival (U-TDOA) is network-based and gets timing messages from uplink signals received from a mobile station at a number of neighboring base stations. LMUs are distributed such that several of them can receive the measurement sequences from any mobile station and calculate the TDOA values necessary for estimating MS position (Figure 9.5). Clock precision and computation power is higher in the fixed units that are involved in the measurement procedure than available in mobile handsets, and many of these units, even up to 50 [2], may contribute to the time difference values. U-TDOA can deliver consistent accuracies within 30m to 50m in different geographically areas, and has proven suitable for meeting the FCC E-911 requirements. A prominent advantage of U-TDOA is that it works with unmodified, or legacy, handsets.

9.7 Angle of Arrival

The angle of arrival (AOA) method is practical for consideration only in network-based applications, as it requires directional antenna arrays for its operation. Even then, large-scale adoption cannot be expected because of the expense of replacing or overlaying installed second and third generation cellular base station antennas. However, the technology is particularly attractive in particular circumstances and environments. Whereas three and preferably more base stations are needed for TDOA arrangements, only two base stations with directional antennas are needed for AOA. Rural regions, where cells are large and often no more than two base stations may hear a mobile terminal, AOA can be a usable solution. Also, line-of-sight paths from mobile to base stations are imperative, and urban areas may be very unsuitable for AOA. AOA could be combined with a distance measuring method, such as TOA, to give a unilateral positioning solution. The deployment of smart antennas in third and fourth generation cellular networks is expected to increase the interest in AOA for cellular positioning.

9.8 Received Signal Strength and Pattern Recognition

Received signal strength data, used on its own, is not useful for cellular positioning for most applications. However, when a database can be put together for a defined region, such as that described in Chapter 6, satisfactory accuracy can be obtained. The system can be handset-based or network-based. In the former case, RSSI readings obtained from several base stations are forwarded from the MS to a location server for database search and comparison. Particularly in the case of a network-based system, the database can contain, in addition to received signal strength, information pertaining to other signal and channel parameters including channel impulse response or another representation of multipath components at a particular location. A unit of information in such a database is called a fingerprint. A fingerprint is unique for a particular small area, referenced to the closest base station, and a positioning system based on a signal fingerprint comparison method could be unilateral. A location method that has been developed and that can be applied to any cellular network is called database correlation method (DCM) [7]. The major task in a pattern recognition system such as DCM is to compile and maintain the database. The compilation is done either by direct measurements at locations throughout the coverage area, by calculations based on known topography, or both. The method has the potential of giving usable accuracy in places where the use of other methods of positioning gives poor results, where there is no line-of-sight path for example, common in urban and indoor locations.

9.9 Problems and Solutions in Cellular Network Positioning

Application of specific positioning technologies often depends strongly on the type of cellular network involved. The bandwidth of the cellular signal determines to a great extent the precision that can be attained in time of arrival measurements, the degree of fading, and the effects of multipath propagation.

9.9.1 Narrowband Networks

Both the analog Advanced Mobile Phone System (AMPS) and the U.S. Digital Cellular standard (USDC) have a limited bandwidth of 30 kHz. A system based on an overlay of digital receivers connected to existing base station antennas was developed using the TDOA method and sophisticated processing for correlation of control channel signals [8]. Control channel messages contain time stamps so that copies originating at different receivers in the vicinity of the mobile unit to be located can be correlated in order to produce the time difference data needed for TDOA positioning. Doppler shifts are also detected in the signals facilitating MS tracking by estimates of the MS speed and bearing.

In order to alleviate deep fading that is characteristic of systems involving narrow bandwidth mobile stations, space diversity antennas are used at base stations. AOA is also used to reduce multipath and provide a supplementary method to a TDOA system to improve location accuracy [8].

9.9.2 CDMA

In contrast to AMPS and TDMA, CDMA has a wide bandwidth, 1.25 MHz, and therefore can potentially make higher accuracy TDOA position estimates. However, network based TDOA requires that several base stations monitor the mobile during TOA measurements, and this is a problem in CDMA [4]. Adjacent CDMA cellular links operate on the same frequency channel, and in order to reduce cochannel interference, strict power control is used. Thus, a mobile that is operating relatively nearby to its serving base station reduces its power to the minimum required for successful communication. The other base stations that are considerably farther away may not be able to hear the mobile in order to make the time of arrival measurements. If the power control is disabled during position measurements, communication links with other mobile units in the vicinity could be impaired. One way to get around the power control problem is for the base stations to do the TOA measurements during soft handovers of the MS between base stations. During handovers, the mobile transmits maximum power. In a position measuring procedure, the control entity can force handovers between chosen base stations and therefore increase hearability among a number of base stations so that TOA data can be collected.

9.9.3 GSM

The bandwidth of a GSM signal is 200 kHz, making it potentially more accurate than AMPS or TDMA for TDOA positioning. However, GSM uses slow frequency hopping as well as TDMA, so coordination among base stations in a network-based system is necessary so the TOA measurements can be made at multiple base stations, each of which knows the mobile's transmission frequency. The use of a common frequency control channel over which to position measurements are made can alleviate the coordination problem. The use of the control channel by the mobile can be initiated by issuing a handover command.

9.10   Handset-Based Versus Network-Based Systems

Network-based systems are preferable to system operators that must comply with the FCC regulations, since they are designed to accommodate legacy handsets. A disadvantage is that networks using these systems may experience reduced capacity, particularly if they utilize the positioning capabilities for popular location-based services, other than emergency. Also, network based location involves a significant infrastructure change and investment, particularly on existing networks that were designed for telephone service which has conflicting requirements compared to positioning functions. Third generation networks that were conceived with positioning requirements taken into account are more amenable to including location functionality than previous generations.

Handset-based solutions have the drawback of requiring special, more expensive and higher power consuming mobile phones. However, they do have advantages. A GPS phone, including A-GPS, is independent of the cellular network from the point of view of positioning accuracy. The accuracy of network-based positioning depends on network-based factors including cell size and density, and in the case of TDOA based technologies, the location of the several base stations within hearing of the mobile unit. If these base stations happen to be located on a straight line, for example, there will be severe dilution of precision (DOP) that degrades accuracy. Even non-GPS handset-based solutions have advantages. They can use downlink control channels, which are not power controlled, and have known frequencies for making TOA measurements from multiple base stations. Technologies that are not network coordinated can also monitor and measure signals from base stations that are not in the same communications network as the mobile, for a wider range of hearable fixed sources. Of course such handset-based systems do have to have transmission time references for all base stations as well as their location coordinates. This information could be supplied by nonnetwork stations located in the vicinity where the mobile is operating. Also, the fact that the handset-based systems do not impinge on network capacity allows them to make continuous position measurements when a location fix is requested, and thereby to increase the positioning accuracy.

9.11   Accuracy Factors

There are several reasons for discrepancies between reported position and actual position of a mobile station target. The network may be required to report to the location service requestor an estimate of the position accuracy. The following factors affect the accuracy of the position measurements [5]:

- Geometric dilution of precision (GDOP) (see Chapter 7, Section 7.3.3);
- Capabilities of signal measuring hardware;
- Effects of multipath propagation;
- Effects of timing precision and accuracy of synchronization.

An accuracy zone may be reported as lengths of major and minor axes of an ellipse around the position estimate, as well as the orientation of the ellipse [5].

9.12   Conclusions

The form of the technical implementation of location in cellular telephony was influenced to a large extent by the FCC regulations concerning emergency calls. Those regulations effectively gave a priority to a network-based location solution, which insures inclusion of legacy handsets and therefore eases the fulfillment of the need to include most of a carrier's subscribers. However, handset-based location methods have an advantage in commercial location applications that favor roaming and decreased network dependence and can give increased precision, although at a higher price. The form of the location solution is also influenced by the particular cellular multiplex technology, TDMA or CDMA.

First and second generation cellular networks present particular problems because they were not conceived with location applications in mind. Third generation cellular networks have protocol and organizational provisions for network-based location services and additionally, their wider bandwidths in comparison to earlier generations should enhance the attainable positioning accuracy. The commercial viability of third and higher generation cellular systems depends on provision of services beyond simple two-way voice connections, and it is to be expected that location services will be widely promoted.

While all methods of wireless location are used in cellular networks, it appears that TDOA predominates for network-based implementations. GPS for handset-based location solutions can give high accuracy, but for low power consumption and high sensitivity needed for useable indoor performance and coverage in urban environments, they are dependent on the wide deployment of satellite acquisition assistance centers.

References

[1] FCC Regulations Chapter 47 Part 20, "Commercial Mobile Radio Services," Section 20.18, amended, 2004.

[2] "E-112 Issues and Answers: Recommendations and Insight for the Optimal Planning and Implementation of E-112, Emergency Wireless Location for the European Union," http://www.findertog.com/e112_issues_and_answers.pdf, 2004.

[3] Gibson, J. D., (ed.), *The Mobile Communications Handbook*, Boca Raton, FL: CRC Press, 1996.

[4] Caffery, J. J., Jr., and G. L. Stuber, "Overview of Radiolocation in CDMA Cellular Systems," *IEEE Communications Magazine*, April 1998.

[5] Technical Specification 3GDPP TS 25.305 v7.0.0 (2005-06), "3rd Generation Partnership Project; Technical Specification Group Radio Access Network; Stage 2 Functional Specification of User Equipment (EE) Positioning in UTRAN (Release 7)," 2005.

[6] Drane, C., M. Macnaughtan, and C. Scott, "Positioning GSM Telephones," *IEEE Communications Magazine*, April 1998.

[7] CELLO Consortium (http://telecom.ntua.gr/cello), "Cellular Network Optimisation Based on Mobile Location," Document Id: CELLO-WP2-VTT-D03-007-Int, November 5, 2001.

[8] Reed, J. H., et al., "An Overview of the Challenges and Progress in Meeting the E-911 Requirement for Location Service," *IEEE Communications Magazine*, April 1998.

CHAPTER 10

Short-Range Wireless Networks and RFID

The wireless networks that are discussed in this chapter have several notable characteristics that are significant from the point of view of wireless positioning. They all operate using low power over relatively short distances, generally no greater than 100 meters. Consequently, for time of arrival methods, time of flight is short, and for useful accuracies, time resolution should be around 10 nanoseconds, equivalent to three meters range. Also, all of the networks are used basically indoors, subjecting wireless signals to nonline-of-sight and severe multipath conditions. These systems all operate on the so-called unlicensed bands, and therefore are susceptible to interference from a wide range of signal types. Mobile terminals are small and must have very low current consumption to extend battery life. They also serve many consumer applications and are designed for low cost. Technologies for incorporating location services in these networks must take into account all of the characteristics mentioned above.

The positioning methods mostly used for the short-range wireless networks are TOA, RSS and fingerprinting. GPS is also used but will not be discussed here. Proximity detection for location is also popular but this can be considered a form of RSS. RSS may be more accurate for short distance position estimations since propagation more closely follows a deterministic rule. TOA is inherently a more accurate estimate of the distance between two terminals than RSS. In a multipath environment, the deviation of the distance estimate from the true value depends directly on the length of the nondirect path. However, when basing the distance estimation on a function of the signal strength, large deviations of the estimate from the true value can occur when the nondirect path differs from the direct path by as little as one half wavelength. In this case, the two signals arrive at the receiver out of phase and the composite signal can be even 20 dB below the value of RSS if only the direct path existed.

Fingerprinting can achieve higher accuracy than direct calculation of distance from RSS. Its biggest drawback is that it requires the creation of a database that is individual for each site to be covered, and whose size depends on the area of that site. Changes in physical details of the site are apt to require updating the database, and transient changes, such as the movement of people at the site, may affect accuracy. Variable antenna patterns due to random terminal orientation will also affect the positioning results. In spite of the problems, there are several short-range location systems based on this method on the market that apparently function satisfactorily for their intended use.

241

The ability to detect multipath in a received signal depends directly on the signal bandwidth. One way to separate and observe multipath signals is to use several correlators with independent time reference signals. The base of the correlation function is approximately two symbol periods. Paths having time delays that differ from the direct path signal and other multipath symbols by more than one symbol period can be distinguished. Indoors, multipath lengths are relatively short, so symbol periods, or chip periods in the case of DSSS, must be short and symbol/chip rate relatively high in order to resolve the multipath signals. While TOF estimating precision can be improved by statistical and processing methods, the multipath phenomena can still prevent good enough distance estimations from being achieved.

Most existing short-range wireless network standards were developed without defined procedures for positioning. New standards and updates of the old ones are taking the need for location services into account.

10.1 WLAN/WiFi

There are many reasons for incorporating location facilities in a WiFi network. A common one is for security enhancement. An intruder can access a network that is installed in offices from outside of the building using directional antennas, high power, and a sensitive receiver. An access point that has a location capability can deny access to such an intruder on the basis of his location, even though signal strength may be at a similar level to legitimate clients in the network. Other uses of location facilities are person and equipment tracking.

10.1.1 TOA

One-way TOA distance measurement requires that the measuring receiver, which estimates the arrival time of a signal from a distant transmitter, knows the time of transmission of that signal in terms of its own clock. For this, the receiver and transmitter clocks must be synchronized to the degree of accuracy required of the distance measurement. WiFi standard IEEE 802.11 specifies the clock synchronization by way of a timing synchronization function (TSF) timer that counts in increments of microseconds. The mechanism described in the standard is designed to maintain synchronization of all TSF timers in a network to within 4 microseconds, plus propagation delay. The timer precision, 1 microsecond, is equivalent to a distance of 300 meters, which is far too large for direct use in the network, where terminals are normally separated by only tens of meters. However, when two-way TOA is employed, transmit and receive timer synchronization is not necessary. In this case, the initiator/interrogator transmits a message to a second terminal, which responds with a return message. The two-way TOA is then the time difference between the period between sending a message and receiving a reply at the initiator, and the period between receiving the message and sending the reply at the responder.

What makes two-way TOA distance measurement a possibility in WiFi (IEEE 802.11) networks and ad hoc connections is that a message response after a fixed time interval from message reception may be automatically generated in terminal hardware, as part of the access protocol. A short description of the part of the protocol that is relevant for a distance measurement routine will be helpful for the explanation that follows. The fundamental access method of IEEE 802.11 is known as carrier sense multiple access with collision avoidance (CSMA/CA), which is used both in ad hoc peer to peer connections and access point (AP) coordinated infrastructure configurations. A station that has a message to transmit and wants to gain access to the channel must assure that the channel is clear before transmitting. It does this by monitoring the channel and only if it is clear for a required period of time can it consider attempting access. The probability of clashes between transmissions of two or more terminals is reduced by a procedure governing random back off, or additional waiting periods, once a terminal finds that the channel appears to be free. An exception to the channel monitoring requirement before transmission is made in the case of acknowledgment (ACK) messages. The purpose of the ACK is to inform the sender that his message was received correctly. If he doesn't receive the ACK during a given period of time after his message was sent, he can try to send it again, while observing the prescribed collision avoidance routine. The message-ACK feature greatly reduces the chance of losing data, even in dense networks and difficult RF transmission channels.

Since ACK is sent blindly, without checking whether or not the channel is busy, the protocol has provisions for protecting it from possible collisions. One aspect of this protection is by assigning to the ACK a time period, called short interframe space (SIFS) between the end of the received message and the beginning of the ACK transmission. SIFS is shorter than the minimum clear channel period before a terminal starts transmission, so that another terminal wanting to access the channel will hear the ACK frame and will postpone its transmission according to the back off rules. Another measure for avoiding collision is effective when a terminal is in range of a message sender but not of the responder. The message, or data frame, contains a duration field that indicates the length of the message plus the lengths of the subsequent expected ACK and a SIFS interval. Terminals listening to the channel, which may not be in range of both sides of a transaction, update a network allocation vector (NAV) variable with the contents of the duration field. Those terminals may not try to access the channel for the duration of the NAV, even when the channel appears to them to be clear.

Time of arrival distance measurement depends on precise determination of time at a specific epoch of a received packet. Three types of IEEE 802.11 physical layer frame formats are shown in Figure 10.1. They all have a preamble that facilitates bit synchronization of the received frame and includes a start frame deliminator (SFD) or equivalent [Figure 10.1(c)], which indicates the beginning of the frame header. The beginning of the header may be the best time to refer to as time of reception of the frame, although any other point could be used by counting chip or symbol periods. Any point on a transmitted frame can be used for reference, since its time is determined by the transmitter's known clock phase.

The two-way TOA principle is illustrated in Figure 10.2. A frame sent from the initiator WiFi terminal to a responder arrives after a propagation delay of T_P. If the frame is received correctly, an acknowledgement transmission is sent back to the initiator. The time difference between the end of the received message and the beginning of the acknowledgement, ACK, must be less than SIFS, which in

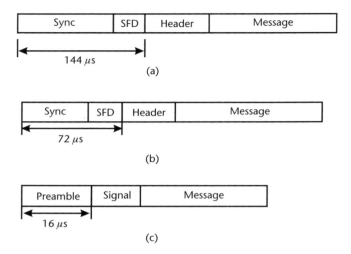

Figure 10.1 IEEE 802.11 physical layer frame formats: (a) long format, (b) short format, and (c) IEEE 802.11a format.

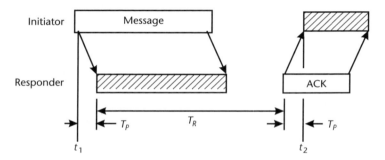

Figure 10.2 Two-way TOF process in WLAN.

802.11b and g has a nominal value of 10 μs. This time includes hardware and processing delays plus the changeover time between transmission and reception. The ACK is sent automatically by chip hardware. The initiator receives the ACK after a propagation delay of T_P. From Figure 10.2 it is seen that the total propagation time is:

$$2T_P = (t_2 - t_1) - T_R \qquad (10.1)$$

In Figure 10.1 the marking of the beginning of a message or ACK frame is the end of the preamble. As seen in Figure 10.3, T_R equals the sum of the total data message frame length minus the preamble, the SIFS, and the preamble length. The accuracy of the estimation of T_P depends for the most part on the precision of SIFS.

There are several sources of uncertainty that make it difficult to get desired accuracy when conforming to the IEEE 802.11 standard. The times t_1, t_2, and T_R when measured in the physical layer have no better precision than that of the chip clock. An example of a chip that implements IEEE 802.11 is Intersil HFA 3863.

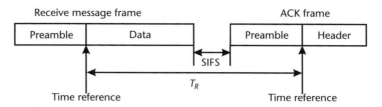

Figure 10.3 Data frame plus ACK.

Its clock source has a frequency of 44 MHz, with a period of 22.7 ns, equivalent to a distance of 6.8 meters. The SIFS is nominally 10 microseconds and can vary in different implementations. This is perhaps the largest source of uncertainty. Inaccurate time bases and the fact that the transmitting and receiving clock rates are often not synchronized also adds inaccuracy that is a function of packet length. For example, let's assume that the receiver clock is synchronized to the transmitting clock through phase locking the input signal with the local oscillator during the packet preamble. If the clock frequency differs from nominal by 20 ppm, and if the true value of $t_2 - t_1$ is, say, 300 microseconds, then the error in the measurement due to clock inaccuracy is

$$\Delta T_{error} = 300 \ \mu s \times 20 \times 10^{-6} = 6 \ \text{ns}$$

For a speed of propagation of 3×10^8 m/s the range error is 1.8 meters.

While the time interval $(t_2 - t_1)$ is measured on the side of the initiator, the most accurate estimation of T_R would have to be determined on the responder side of the link. This is because the responder determines, in hardware, the SIFS. When the responder measures T_R, it could transmit the estimation result back to the initiator for calculation of $(t_2 - t_1) - T_R = 2T_P$. However, in many cases, this is undesirable or impossible. For example, if the distance estimation is intended to be used for security against unwanted network access, there would be no cooperation from the intruder. However, the initiator can approximate T_R. He knows the message length and the ACK preamble length and can assume a SIFS time to be the same as at the initiating terminal, in which case it will be exactly cancelled. If this were the case, the propagation time estimate would be, from (10.1) and Figure 10.2:

$$T_P = \frac{1}{2} [(t_2 - t_1) - (T_{DATA} + SIFS + T_{PREAMBLE})] \qquad (10.2)$$

An example of a WLAN time of flight locating system uses special tags and standard hardware APs. A high precision clock sets the SIFS in the tag, so that the interrogating AP calculates T_P from (10.2) when knowing the measured values t_2 and t_1. Assuming SIFS is known exactly and that the receiving path is line of sight, T_P can be found after one message-ACK sequence to a precision approximating the AP clock resolution.

Useful accuracy in estimating range can be obtained without high precision clocks by exchanging the clock precision for measurement time [1–3]. The range accuracy using the time of arrival method is a function of the number of measurement samples, the actual range, and also the length of the message, or bit period times a given number of bits. The results of measurements reported in [2] showing range resolution as a function of number of measurements are presented in Table 10.1. The data was taken with a nominal duration of 160 microseconds from message transmission to ACK reception, a message data rate of 11 Mbps, and a confidence level of 99% over ranges from 0 to 100 meters. An Intel Prism chip set was used, with added hardware to access pins where t_1 and t_2 (Figure 10.2) can be measured. Timing resolution was 22.7 ns from the 44 MHz clock. Assuming 10 milliseconds for each measurement, the 871 measurements reported for a resolution of 10 meters would take 8.71 seconds to accomplish.

Useful ranging accuracy has been demonstrated with no hardware modification but with special software to use the 802.11 time stamp resolution of 1 microsecond [1, 3]. Measurement error of t_1 and t_2 and the unprecise duration of SIFS, together with the basic timing resolution of 1 microsecond for time stamps of received messages for some IEEE 802.11 chip sets, causes repeated measurements of $(t_2 - t_1)$ and T_R [measured as $(t_2 - t_1)$ at zero range] to change between step values that differ most of the time by 1 microsecond [1]. The random phenomenon that can explain the jumps between two discrete values is Gaussian noise associated with the measurement, but even if the noise is too low to cause the changes, another effect can explain the jumps, according to [1]. This is the relative clock drift between the two terminals due to slightly different frequencies of the crystal controlled reference oscillators on the circuit boards. The relative clock drift causes rounding errors in the timing measurements between the sent and received frames. The result is that even though basic timing resolution is 1 microsecond, much higher resolution can be obtained by averaging the time differences over a large number of trials.

The way averaging of random discrete time differences increases resolution is demonstrated by the rough histogram of Figure 10.4. Let T equal the initiator measured time interval, $T = (t_2 - t_1)$. Accounting for the message length and propagation delay, we let the true time interval be $T_{TRUE} = 150.25$ microseconds. Measurement noise, caused by thermal noise and relative clock drift, makes the results of multiple measurements of T, for which the resolution is 1 microsecond, be either 150 or 151 microseconds. The unbiased sample average (error due to absolute clock drift is ignored for simplicity) T_{AV} equals T_{TRUE}. $T_{AV} = 150.25$ microseconds is indicated by a dotted vertical line in Figure 10.4. Figure 10.4 shows as a dashed curve the continuous Gaussian probability density, for a standard deviation of 0.3 microsecond. The measurements are rounded off to the nearest

Table 10.1 Range Resolution Versus Number of Measurements

Range Resolution (m)	Number of Measurements
10	871
15	388
20	218
25	140
30	97

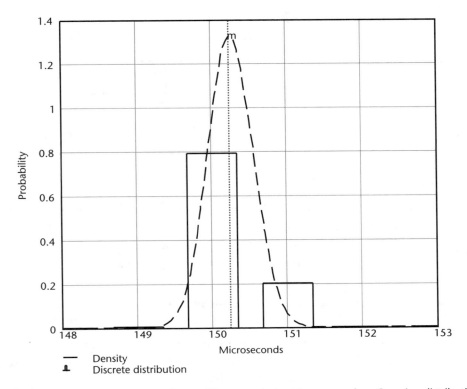

Figure 10.4 Discrete distribution of time differences derived from a random Gaussian distribution.

microsecond and the solid outline bars show the probability of a measurement result being 150 or 151 microseconds. If 200 measurements of t_1 and t_2 are taken, 160 of them will give $T = 150$ microseconds and 40 will result in $T = 151$ microseconds, corresponding to the probabilities shown in Figure 10.4 of 0.8 for $T = 150$ microseconds and 0.2 for $T = 151$ microseconds. Averaging the results of all measurements gives $T_{AV} = 150.25$ microseconds. Continuing the example to get the range estimation, let $T_R = 150.17$ microseconds, found by averaging multiple measurements at zero range. Then $T_P = (1/2) \times (T - T_R) = 0.08$ microsecond. The distance between the two terminals is $0.08 \times 10^{-6} \times 3 \times 10^8 = 24$ meters.

Results of range estimations based on the time stamp resolution of 1 microsecond are reported in [3] based on averaging over a large number of measurements. For example, average estimated distances compared to actual distances are presented in Table 10.2 from 10 to 50 meters and a data rate of 24 Mbps.

The results in Table 10.2 were obtained by taking 4,000 measurements over a time of 17.5 seconds. Time stamp resolution was 1 microsecond.

These are several advantages of using two-way TOA over other methods:

- Even with one AP, range can be determined to an accuracy of a fraction of a meter in devices having high precision clocks.
- Unambiguous two-dimensional location can found from triangularization using a minimum of three APs. Simultaneous time measurements by the APs are not needed and there is no synchronization requirement.

Table 10.2 Estimate Versus Actual
Distance Obtained Using 4,000
Measurements Per Estimation

Actual Distance (m)	Estimated Distance (m)
10	10.4
20	19.4
30	28.9
40	39.1
50	49.0

- Using averaging, TOA resolution far better than that of the local clock is attainable.
- Knowledge of radiated power is not needed.
- Host (AP) or client range initiation is possible.

Two notable limitations to two-way TOF ranging are:

- Multipath in an indoor environment reduces accuracy.
- Relatively long measurement time is required for high accuracy when a high resolution clock is not available.

10.1.2 TDOA Methods for WLAN Location

Using an infrastructure designed or adopted for the purpose, TDOA can be used to locate mobile stations. In order to get a TDOA fix, at least three APs measure TOA at the same reference point on a frame transmitted by a mobile station. The station is associated to only one access point at a time, so the additional APs participating in the measurement must be within hearing range of the station and must know which frame to measure. The location server issues a command to all APs in the network to find a particular wireless device. Clock precision must be commensurate with the desired position resolution and the clocks of participating APs must be synchronized. The identification task is simplified if an ACK frame from the target is used, since it is transmitted immediately following what can be a measurement triggering frame from the associated access point. Location is found by getting the least mean square coordinates of hyperbola intersections when the positions of the participating APs are known.

TDOA positioning has advantages over two-way TOF. When high precision clocks are used, the location fix is relatively fast, and the target movement can be tracked over reasonable target speeds. Measurements are made without the participation of the target and are virtually impossible to spoof. A disadvantages is the need for clock synchronization of APs, and of placing APs so that three or more will be in hearing range of targets that are positioned anywhere in the desired coverage area. Nonline-of-sight propagation will degrade accuracy, but the participation of multiple APs in the measurement will improve the location result.

10.1.3 Fingerprinting

Fingerprinting is the most widespread positioning technique used for WLANs. It is based on a set of RSS measurements taken of a target from multiple access points and comparison of the results with a previously compiled database (see Chapter 6). The method has several advantages. No time synchronization is required, as for TDOA. The reading of RSS is inherent to the IEEE 802.11 protocol, and no special hardware is needed. Tags or devices based on the IEEE 802.11 standard can be tracked. The method is particularly applicable to indoor networks as the vagaries of multipath propagation are automatically accounted for in the reference database. On the negative side, the method implies creating a database for the area to be covered, and changes in AP deployment and physical features of the environment require updating the database. The method is computationally intensive and a special location server is required for position outputs [4].

The fingerprint database is created from a grid mapped to a floor plan of the coverage area that includes physical characteristics—partitions, wall materials, furnishings—and the position of access points. Ray trace software creates vectors of signal strengths at grid positions throughout the area, and actual measurements are added as needed for increased accuracy. RF signal strength prediction is based on reflection, attenuation and multiple transmission paths between grid points and each AP. Grid points can represent an area as small as 15 cm square [4]. Real-time signal strengths from a target to all access points in range are compared to the database to estimate the target location. Targets can be tracked to an accuracy of a few meters.

Another slightly different approach to the fingerprint technique has been suggested [5]. Separate receive only, or passive, sniffers are installed in the location coverage area. A dedicated infrastructure performs the location function independently of the network. There are several advantages to this approach. The dedicated sniffers scan WLAN RF channels continuously with signal strength and station identity information time stamped for correlation and processing at a location server. Thus there is no bandwidth overhead on the network for the location function. The inexpensive sniffer receivers can be deployed for best positioning geometry, instead of being restricted to AP locations that are selected for optimum communication coverage. The sniffers are entirely passive and communicate with the database and location server through wired Ethernet. Emitters may be added to the location infrastructure to facilitate regular and convenient database updating, particularly applicable to a dynamic environment. The independent sniffers can be deployed for minimum database profiling, which means a relatively low number of database points for a given area and location performance.

Reference [5] also studies a client based location method, in addition to the sniffer approach. In a client based system, a target takes RSS readings from a number of APs in an area and compares them to a database that was created in a similar manner, that is, from client RSS readings. We have referred to such a system as a unilateral one. The client, or target, measures signal strengths from different access points using probe request responses.

Figure 10.5 shows median error versus number of profiling points for a client based system and a sniffer based system [5]. The site was office space with an area

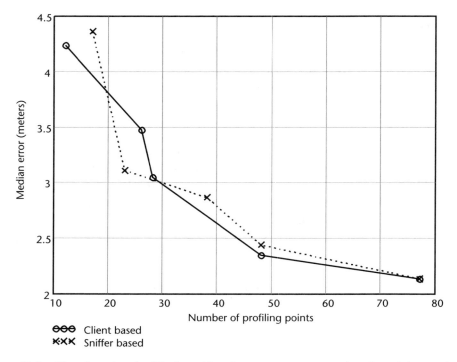

Figure 10.5 Client-based and sniffer-based location system errors as a function of the number of profiling points. (*From:* [5]. © 2004 IEEE. Reprinted with permission.)

of 32,000 ft^2 (3,000 m^2) in which five sniffers colocated with five access points were deployed. The results for the two systems are comparable. For example, a database with profiles of 60 test points within the area can produce a median error of better than 8 feet (2.4m). Each method has its own advantages. The client based system uses the existing WiFi network with no additional infrastructure. However, it cannot be used to locate rogue or otherwise unwanted terminals in the network. The sniffer method has the advantages explained above.

Technologies based on the RSS fingerprinting method of location predominate for WLAN networks. They are particularly effective for indoor use because they are not affected by multipath propagation. The biggest disadvantage is the necessity for creation of a database that is specific for a particular installation. Accuracy is generally adequate for most applications but could be reduced by temporary physical obstructions or deviations in client antenna radiation patterns, a problem with small, portable devices. Another strong factor in favor of the fingerprint method is that it may be based entirely on software with no hardware changes to legacy client stations or access points.

The TOF method of distance measurement has the potential of achieving higher accuracy than technologies based on RSS, it does not have to be adapted to each installation site by preparing a database, is not sensitive to antenna orientation and is only marginally affected by blocking. However, accuracy is strongly affected by multipath propagation and high measurement rate and precision require dedicated hardware, at least for terminals on one side of the communication link. The bandwidth of WLAN systems based on IEEE 802.11—22 MHz—is too low to extract the direct path from reflections in an indoor environment [6]. The rise in the use of smart antennas which combat multipath interference may bring with it greater use of TOF in location systems for WLAN.

10.2 WPAN

While IEEE 802.11, or WiFi, dominates WLAN, there are several modes of operation and standards that make up WPAN—wireless personal area networks. These networks don't have the infrastructure that is part of many WLAN applications and range is generally much shorter, nominally 10 meters. WPAN standards differ significantly in their physical layers. Principal characteristics of WPAN networks are shown in Table 10.3.

In particular, standards IEEE 802.15.4a and ECMA-368 include positioning features. The most basic form of position information from a WPAN client is to note the location of the host station. This proximity method gives an accuracy in the range of 10m. Significantly higher accuracy, then, must be around 1m, although any improvement could be useful for some applications. Achieving 1-m accuracy is possible for the systems with high data rates and high bandwidth. UWB is particularly suitable for high precision ranging.

10.2.1 Bluetooth

A method of making coarse estimates of Bluetooth device location uses an approximation of the maximum range of a link [7, 8]. Bluetooth position servers are located throughout an area where a client position is to be estimated. These servers are programmed to give their location coordinates to a requesting client. Bluetooth devices that do not have special programming to handle a position request can also be used. In this case, the position of the device is contained in a table referenced by the fixed device's ID that can be accessed by a client from a special location server that has been set up for this purpose. The specified power class 3 has a maximum power output of 0 dBm which enables a range of around 10 meters. The positioning method assumes that if a connection is made, the client device is within 10 meters of the server. Higher accuracy is obtained as shown in Figure 10.6 when the client makes a connection with two or more position servers, or

Table 10.3 Characteristics of WPAN Networks

Description	Standard	Raw Data Rate	Modulation Scheme	Frequency Band
Bluetooth	IEEE 802.15.1	1,3 Mbps	FHSS	2.4 GHz
Hi rate WPAN	IEEE 802.15.3	11–55 Mbps	PSK, QAM	2.4 GHz
Hi rate alternative PHY	ECMA-368	53.3–480 Mbps	UWB	3–10 GHz
Low rate WPAN (ZigBee)	IEEE 802.15.4	250 kbps*	DSSS	2.4 GHz*
Low rate alternative PHY	IEEE 802.15.4a	850 kbps (UWB user rate)	Chirp, UWB	2.4 GHz, Chirp 0.5, 3–10 GHz UWB

*Also specified for 868–870-MHz and 902–928-MHz bands at 20 and 40 kbps, respectively.

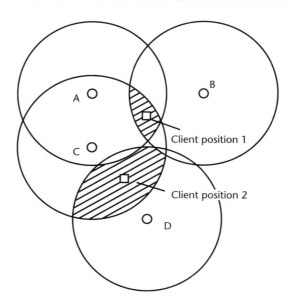

Figure 10.6 Bluetooth positioning by triangulation with constant radius references.

devices whose locations are in the location server table. In Figure 10.6, the client at position 1 has established connections with devices A, B and C whose coordinates are obtainable. Device D is out of range. The client is located in the shaded region. Estimated position coordinates can found by calculating the centroid of the region, or an approximation from the intersections of the constant radius circles that bound the region. When the client moves to position 2, it is in range only of devices C and D, and is located in the corresponding shaded region.

In order to make location estimations, a mobile client must attempt to connect to several Bluetooth devices and then to receive reference coordinates from those with which the connection was successful. The average time to make a location estimation when five location source devices were queried was 19.2 seconds, and calculated worst case was 31.3 seconds [7]. These times are long compared to the time, 16.7 seconds, it takes a client to pass through the 20-m diameter of the range circle at an average walking speed of 1.2 m/s. Therefore, accuracy of the method may be seriously impaired when position measurements are taken when the client is in motion. Also affecting accuracy is the fact that maximum range may vary widely from the assumed 10 meters. The radiated power and antenna pattern of hand held devices varies according the way they are held and their distance from the body, and indoor transmission path obstructions limit the accuracy of the presumed range.

In spite of its indefinite accuracy, the method described is still an improvement over a system that estimates its position purely on the known location of one device in a link, that is, a proximity system. It requires no hardware modifications or changes to Bluetooth protocol, and may be adequate for certain location services.

Bluetooth positioning using the RSSI measurements reported in the Bluetooth physical layer has been investigated [9]. In addition to the difficulties of relating signal strength to distance in an indoor environment due to reflection caused fading and shadowing, the RSSI measurements in Bluetooth are imprecise and do not relate in a linear manner to the strength of the received signal. The prime purpose of the RSSI reading in Bluetooth is to provide feedback for the power adjustment of the transmitter to which the receiver is linked. When the received signal is within a 20 dB range between an upper threshold (strong signal) and a lower threshold (weak signal), called the golden received power range, the reported RSSI is 0. The lower threshold is somewhere between 6 dB above the actual receiver sensitivity and −56 dBm. The upper threshold is 20 dB greater, ±6 dB. The RSSI is negative when the signal is below the lower threshold, indicating that the transmitter should increase power, and positive when the signal is above the upper threshold, causing the opposite transmitter to decrease its power. The RSSI output versus input signal strength characteristic of Bluetooth is shown in Figure 10.7 [10]. The curves in the figure are based on four different Bluetooth devices and give an indication of the spread among different Bluetooth modules. One approach for increasing the input range over which RSSI can measure input signal strength is to use variable attenuators in separate antennas that can be switched in to a Bluetooth access point [10]. The attenuators shift the linear portion of the RSSI curve so as to facilitate measuring signal strength over a wider range.

Measuring RSSI to obtain link distance to several Bluetooth access points at known positions and then using triangulation calculations to estimate the client coordinates has been used to obtain an average error of around 2 meters in tests [9]. A range of up to 8 meters was possible using power class 2 devices with 4-dBm output. The range was necessarily reduced below maximum possible for communication, so that the received signal strength would be above the upper threshold, causing a positive RSSI reading. For this method, preliminary measurements were made to relate RSSI to range. As reported in [10], RSSI measurements

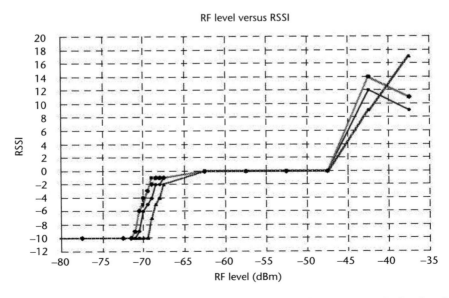

Figure 10.7 Bluetooth RSSI versus signal strength. (*From:* [10]. © 2004 IEEE. Reprinted with permission.)

were taken at random locations in a 46 square meter room, and curve fitting was applied to the averaged results. A log function and two polynomial functions were tried to get the best fit. A graph of RSSI versus range using the log function is shown in Figure 10.8. The corresponding function is

$$RSSI = -2.28 \ln(d) + 5.7 \qquad (10.3)$$

where d is the range in meters. This system does not require any Bluetooth hardware or software modifications but a disadvantage is that a preliminary site survey to relate RSSI to distance is necessary to get reasonable accuracy. Since RSSI precision is relatively low in Bluetooth, a large spread in RSSI versus signal strength could be expected over different Bluetooth devices, further reducing the accuracy.

Much higher accuracy than that obtainable from proximity methods or those using RSSI can be achieved by measuring phase differences of frequency hopping carriers, as described in Chapter 5. Special hardware is required and Bluetooth protocol is not adhered to during the measurement process. However, for proprietary devices, the hopping channel separation of 1 MHz and total span of 80 MHz are appropriate for achieving range precision in the vicinity of 1 meter. The Bluetooth protocol can be adhered to for all operational modes, including connection and normal data and voice communication activities. For distance measurement, two significant deviations from the Bluetooth system are needed. Two-way phase shift due to time of flight must be measured. This requires that the initiating transmission and the reply must be on the same channel. A second requirement is for inhibiting modulation and transmitting a cw carrier on the channel center frequency during a portion of the transmission slot in both directions. An example of the time slots used for distance measuring is shown in Figure 10.9. Implementation of the system is described in Chapter 5.

During the period in the slot of cw transmission, a narrow baseband filter is used, significantly increasing the carrier to noise ratio for accurate phase measure-

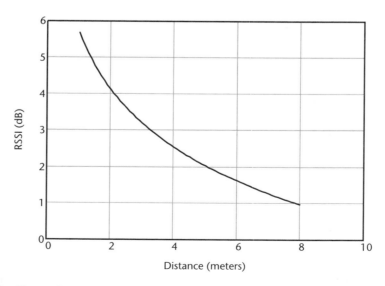

Figure 10.8 Bluetooth RSSI versus distance.

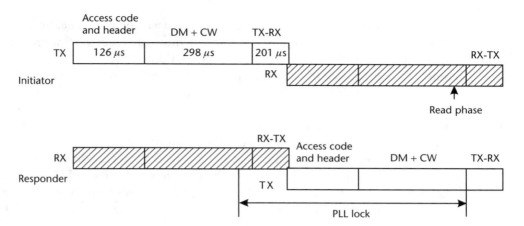

Figure 10.9   Frequency hopping distance measurement slots.

ment. Accuracy is reduced by oscillator drift during the time from responder phase lock on the incoming carrier and the phase difference measurement at the interrogator. However, the biggest detriment to accuracy is multipath reflections. Digital signal processing in the frequency domain can improve accuracy in a multipath environment.

10.2.2   ZigBee

ZigBee is Bluetooth's cousin in the family of short-range low data rate standards for wireless personal area networks (WPAN). Based on the IEEE 802.15.4 standard [11], it is intended principally for monitoring and control applications that demand very low power consumption from a battery source and low price. When used in multiple node sensor networks, some nodes will be static with known locations and others mobile with location capability. Range coverage of a ZigBee node in indoor use is 20 to 30 meters. The most suitable methods of distance measuring and location for IEEE 802.15.4 devices is that based on signal strength indication.

IEEE 802.15.4 defines flexible network topologies. In a star network, one device is a personal area network (PAN) coordinator and other devices communicate only with it. By contrast, devices in a peer-to-peer network topology are capable of communicating with any other network device within their radio range. One device in this network also is designated as the PAN coordinator. IEEE 802.15.4 devices are divided into two classes according to their capabilities—full function devices (FFDs) and reduced function devices (RFDs). RFDs can only connect to FFDs. The architecture of ZigBee is defined such that clusters of devices within range of each other can associate with adjoining clusters and so on, forming cluster chains or meshes that extend far beyond the range of a single device. Within the group of associated clusters, only one device is the PAN coordinator. Such an arrangement facilitates location capability over a wide area.

A ZigBee receiver provides a measured value that can be used for range estimation during the reception of each data message. It is called link quality indication (LQI) and may be based on RSSI. LQI is an 8-bit value that is passed on to higher layers along with the message data and can be used for a distance measuring or location finding program. LQI may be implemented using receiver energy detection, a signal-to-noise ratio estimation or a combination of the two. The specification requires that at least eight levels of LQI be available. The receiver energy detection (ED) measurement is presented as an 8-bit integer. As specified in IEEE 802.15.4, the minimum ED (0) indicates received power less than 10 dB above the specified receiver sensitivity of −85 dBm, and the range of the received power covered by ED is 40 dB or greater. The ED value is the result of averaging over eight symbols, equal to 128 microseconds. Mapping the received power in dB to the ED value must be linear with an accuracy of ±6 dB. The intended use of ED is for scanning frequency channels to find those that are available for a connection, and for clear channel assessment before attempting to access the channel in the CSMA-CA (carrier sense multiple access with collision avoidance) procedure.

While IEEE 802.15.4 does not directly obligate RSSI performance that is usable for distance measurement, many chips do have a suitable capability. When RSSI is related to received power from three or more reference terminals with known position coordinates, a terminal that needs to find its position, known as a blind node, can estimate it by triangulation. In a multilateral arrangement, RSSI measurements are taken at the reference nodes and the data is sent to a control center for position determination. For the unilateral case, the blind node itself measures RSSI from nearby reference nodes, performs position calculations, and sends the coordinate estimates to a controller that uses the information. Each arrangement has its advantages and disadvantages. With multilateral location, the blind node, or target, needs no special capabilities, and may be an RFD that is included, for example, in a very low cost tag. However, the reference devices must coordinate their measurements among themselves, and send each of their data, typically over the network, to the control unit. Thus, the location service activity may be a significant load on the network and is apt to reduce battery life. The unilateral location mode offers reduced network loading and power consumption, but the target device must have the computation capabilities, in hardware or software, to process the RSSI data from the reference nodes to get its own coordinates.

As in virtually all RSSI location methods for indoor environments, the accuracy of ZigBee signal strength positioning can be improved by tailoring parameters used in the calculations to the specific coverage area. The complexity of the propagation model for the area can be reduced by applying statistical analysis to the measured signal strengths. In the presence of fast fading, minimum signal strength is significantly deeper than maximum, and to the steady state signal strength, so peak detection and averaging over a number of readings can improve the received signal strength estimate [12]. This method is applicable only if one of the terminals of the link is in motion, or if surrounding reflecting objects are moving such that multiple signal path lengths change in time. In a purely static situation, multiple readings give the same value, except for measurement noise that is usually much less significant than deviations from the steady state signal strength caused by multipath propagation.

Texas Instruments CC2431 is an example of a ZigBee system on a chip that includes a hardware location engine [13, 14]. The chip calculates an estimate of the chip host's position based on signal strength measurements from three to eight reference nodes and two propagation parameters. The RSSI measured values must be related to the input signal strength Pr_{dB}. In the CC2431 this is done by adding a negative offset constant to the value reported by the physical layer on the chip. The RSSI value as a function of signal strength is approximately linear between −95 dBm and −5 dBm. Distance to the reference mode is calculated according to the following propagation law:

$$RSSI = -(10n \cdot \log_{10} d + A) \qquad (10.4)$$

where n is the path loss exponent, d is the distance between terminals in meters, and A is the value of the average received power in dBm from the reference transmitters at a distance of 1 meter. A depends on the transmitter power and the antenna gain pattern. There is only one value of A per position calculation, so A is common for all reference transmitters, which should have approximately the same power. At 2.4 GHz, if transmitted power is 0 dBm and assuming free space propagation, A averaged in all directions at 1 meter from the antenna equals approximately 40 dBm. n, the average path loss exponent in the area of the location measurements, should be determined experimentally by getting received power values at a mobile device located at various locations throughout the measurement area when distances between the mobile and reference devices are known. An example of the regression curve for path loss versus received power among the measured data points is given in Figure 10.10 [14]. A is the y-axis intercept (not shown in Figure 10.10) and n is the slope of the curve. When d is found for all participating reference nodes, triangulation and an algorithm not specified in [14] is performed to produce an estimation of the coordinates of the node with a resolution of 0.5 meter. Attainable location error is claimed to be better than three meters. Factors serving to reduce accuracy are shading by objects obscuring each reference terminal, multipath, and nonconsistent antenna patterns.

10.2.3 Alternate Low Rate WPAN Physical Layer IEEE 802.15.4a

In 2002 an IEEE committee began working on an alternative physical layer for 802.15.4. A principal objective was to provide high precision and ranging/location capability to 1 meter accuracy or better. The basis for achieving this is an ultra-wideband (UWB) radio operating in the FCC authorized unlicensed band between 3.1 and 10.6 GHz. The distance measuring method will be based on time of flight with timestamp precision to a fraction of a chip duration of around 2 ns. IEEE 802.15.4a will include specific commands (primitives) for providing distance parameters to upper protocol layers.

Distance measurement in a UWB system is based on accurate determination of an epoch in a transmission packet. IEEE 802.15.4a UWB will achieve bandwidth spreading using sequences of narrow envelope pulses. This method is called direct spreading, in contrast to other UWB techniques, such as multiband OFDM. A UWB pulse sequence is shown in Figure 10.11. Correlating a locally generated sequence with the received signal gives an accurate time measurement that can be used for two-way distance measurement or TDOA location positioning. Chapter 11 has more information on UWB and IEEE 802.15.4a.

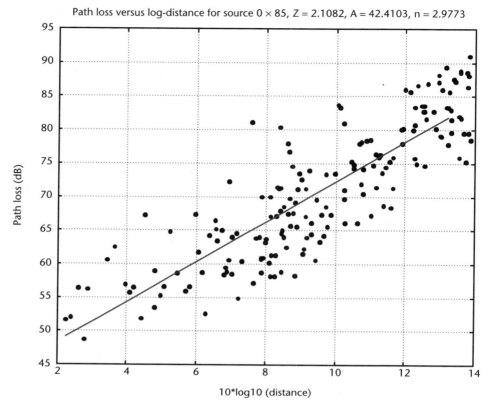

Path loss versus log-distance for source 0 × 85, Z = 2.1082, A = 42.4103, n = 2.9773

Figure 10.10   Regression curve for path loss versus distance. (*From:* [14]. © 2006 David Taubenheim and Spyros Kyperountas, Motorola Labs. Reprinted with permission.)

Figure 10.11   UWB pulse sequence.

10.2.4 ECMA-368 Standard

Standard ECMA-368 specifies a medium access control (MAC) sublayer and a physical layer (PHY) for a high rate personnel area network [15]. It is based on a multiband OFDM technique that was proposed for IEEE 802.15.3a, an alternate PHY for the high rate WPAN standard IEEE 802.15.3. IEEE 802.15.3a was not completed due to lack of agreement on the ultrawideband physical layer technology. Included in ECMA-368 are provisions to support ranging measurements between devices using two-way time transfer techniques. Accuracy of the ranging result is specified to be 60 cm or better. Time stamps are taken from a 32-bit counter that is clocked at 4,224 MHz, with options for clocking at 2,112, 1,056, and 528 MHz. The principle of the ranging operation is shown in Figure 10.12.

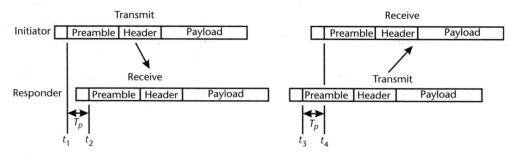

Figure 10.12   Ranging in ECMA-368 [15].

A timing reference point is defined in the specification as the instant when the timing counter is read. It is the instant in a packet preamble at the end of a synchronization sequence when the position of data symbols is accurately known. In the diagram, the initiator takes a time counter reading, t_1, on an outgoing ranging packet. The responder receives the packet after a propagation time T_p and records t_2. A response message is sent from responder to initiator and the transmission time t_3 is recorded. The initiator records the time, t_4, when it receives the packet after another T_p propagation delay. The recorded times t_1 through t_4 are adjusted for delays between the time the signal reached the antenna and the actual time that the counter was read. These delays are a ranging transmit delay, T_{td} and a ranging receive delay, T_{rd}. The responder sends its corrected timing period measurement to the interrogator. The distance D between the two terminals can then be calculated as

$$D = c \cdot \left[\frac{(t_3 - T_{td}) - (t_2 - T_{rd}) - (t_4 - T_{rd}) + (t_1 + T_{td})}{2} \right] \qquad (10.5)$$

where c is the speed of propagation (speed of light). A specified MAC command in the specification instructs the initiator to request from the responder a number of range measurements. Averaging the multiple results improves accuracy but increases the measurement time.

10.3   RFID

Both standard based technologies and proprietary protocols are used for RFID ranging and location. In RFID the target terminal is packaged in the form a tag that is attached to an object whose location is to be tracked. RFID works over a wide range of frequencies—from submegahertz to microwave. An RFID tag may be active or passive. Active tags have internal battery power while passive tags are energized by an external electromagnetic field, generally produced by the tag reader. RFID applications require compact size, low power consumption for long battery life (active tags) or usable range (passive tags), and generally very low cost. RFID systems that provide a location service are often referred to as Real-Time Location Systems (RTLS). Typical RTLS tag devices are active, that is, contain a power source, and must communicate with several spatially distributed readers. An exception that uses passive tags and a single reader is described in Section 10.3.3. IEEE 802.11 and ZigBee devices are particularly popular for RFID location. Most RFID applications, however, are based on proprietary implementations. The three location methods—TOF, RSSI and AOA are found in RFID products.

10.3.1 Proximity Location

The basic method of location using RFID tags is indication of proximity. Either the tag or the reader may be the mobile terminal whose location is to be estimated. In a common example, RFID card or tag readers are installed at defined locations within an office area or industrial facility, often at entrances to rooms and along corridors. People or equipment whose location and movement is being monitored are equipped with RFID tags. When a tag is in the vicinity of a reader, the reader communicates the tag ID to a central location server, either over permanent infrastructure wiring or a wireless network. The location server receives the identity of the reader whose location is known and the tag ID and therefore can log the whereabouts of the person or equipment to whom the tag ID is assigned.

An application where the tags are fixed and the readers are mobile is described in [16]. Passive tags are fixed on articles throughout locations on a university campus. IEEE 802.11b WLAN mobile terminals—laptop or notebook computers carried by personnel whose location is monitored—are equipped with RFID tag readers that sense the identification numbers of the tags in their vicinity. The IDs are reported to a central server over the wireless network that thereby knows the approximate location of the mobile terminal. The accuracy of the location position depends on the particular RFID technology used. Passive tags operating at 125 kHz typically have a range of up to 50 cm, whereas the range of a 13.56 MHz system may be several meters. The system provides a rough location estimate when persons are not near tags using received signal strength (RSS) from the mobile terminals at the WLAN access points.

10.3.2 Distance Bounding for Security

An interesting application for distance measurement in an RFID system is security enhancement by distance bounding [17]. Even where encryption and an authentication protocol are employed, it is possible for security to be compromised. RFID tags or contact less smart cards are used to indicate the presence of the card holder within a short distance of a card reader. A cryptographic authentication routine uses a secret key to verify that only the card with given identity is in the proximity of the reader. However, the system can be fooled by implanting a false, or proxy, smart card near the reader and a proxy reader near the legitimate card, which is outside of the intended operational range for card authorization. The purpose of the system may be to permit entrance to a restricted area or to verify presence of a person, say a security guard, at a particular post. The proxy card and reader will initiate the authentication routine and relay the challenge-response dialog between the real reader and distant smart card, thereby gaining access or asserting a false presence, without having to know the secret key or otherwise breaking the cryptic code.

One way for the reader to be sure that its identified correspondent—a contact less smart card—is within the required bounds of proximity is to measure the elapsed time between message exchanges, that is, detect time of flight. The distances involved in the case of smart cards and readers are tens of centimeters, giving propagation times within a small number of nanoseconds. Reference [17] describes a protocol where challenge and response message units are individual bits. A sequence of exchange of n bits is used for authorization. Reference [17] proposes a method of measuring propagation time, and consequently distance, between the verifying reader and the card or tag whose identity is to be authenticated. Reader and card time base clocks are synchronized by the reader's carrier wave at 13.56 MHz. The message bits themselves are sent over an ultrawideband (UWB) link that has the bandwidth needed for the required time of arrival resolution. A challenge-response bit exchange, with 13.56 MHz carrier zero crossing synchronizing instants, is shown in Figure 10.13. Time intervals t_1 and response delay t_d are known in advance by the system. The sampling time t_s must be found by the reader using a search algorithm to find the pulse peak over several trials, after which the response bits from the card can be recorded. In the UWB exchange, biphase modulation is used where bits are represented by pulses of opposite polarity. Once t_s is found, propagation time is calculated as

$$t_p = \frac{1}{2}(t_s - t_1 - t_d) \tag{10.6}$$

These times t_p are averaged over the complete sequence of bit exchanges and the distance between smart card and reader is found to be

$$d = \overline{t_p} \cdot c \tag{10.7}$$

where $\overline{t_p}$ is average propagation time and c is the speed of light. By defining an upper limit for d, the system can be sure that the authenticated smart card is located within the required proximity of the reader.

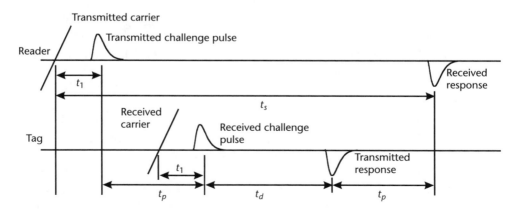

Figure 10.13   UWB pulses synchronized by 13.56-MHz carrier in TOA distance measurement.

10.3.3 Accurate RFID Location

Combined methods of TOF and AOA provide the basis for a location technology for RFID transponders. The operation of one such system, named RFID-radar, is similar to that of radar installations in that is uses a single transmitting terminal, in this case a RFID card reader, to obtain two dimensional coordinates of a target [18]. The AOA function is carried out using the phase interferometer principle described in Chapter 8. Directional reader antenna elements provide gain for reception of signals reflected from the passive transponders and rejection of spurious signals outside of the response region of ±32°.

The system uses passive transponders and operates on spot frequencies within a band from 860 to 960 MHz. The transponder cards absorb operating power from the signal transmitted by the reader and transmit their identity back to the reader by effectively modulating their antenna cross section. Thus, the transponders expend no RF energy and the reader receives modulated reflections of its transmitted wave. Reader transmitter power is between 0.5 and 4 watts and the manufacturer claims a measuring range of several tens of meters.

The distance measuring system uses a bandwidth of 10 kHz and in order to achieve a range resolution on the order of one meter or better, multiple distance measurements must be taken and averaged. An initial position determination for a transponder takes 20 seconds. Thereafter, location is tracked and position is reported at 1-second intervals.

10.4 Conclusions

We have discussed in this chapter two classes of platforms for location applications. One class consists of networks with strict lower layer protocols that generally were not designed specifically for positioning. Examples mentioned were WiFi, Bluetooth, and ZigBee. The other class, RFID applications, has proprietary link protocols that may be created specifically for distance measurement and location.

Short-range location applications are in most cases intended for indoor use and therefore must contend with severe multipath and shadowing impairments. On the other hand, the design of the devices involved have restrictions on cost, size and power consumption, which precludes achieving high accuracy through fast clock rates, accurate RSSI, or sophisticated processing algorithms. When tracking of mobile targets is involved, resulting in random unbiased measurement results, averaging over a large number of samples provides suitable accuracy at the expense of measurement time.

All of the basic distance measurement and location methods—TOA, RSS, and AOA—have been applied to short-range wireless systems. We have seen that TOA is particularly used in WiFi networks, while examples of RSS use in Bluetooth and ZigBee were described. AOA is less prevalent for short-range location applications but may be appropriate for tracking RFID tags.

The location function is being incorporated in new network standards and alternate versions of present standards. Notable examples are the low rate WPAM standard 802.15.4a and high rate UWB standard ECMA-368. Incorporation of distance measurement features in the link layer hardware will give an impetus to the development of low-cost wireless products that have higher positioning accuracy and short measurement time.

References

[1] Gunther, A., and C. Hoene, "Measuring Round Trip Times to Determine the Distance Between WLAN Nodes," *Proc. Networking 2005*, Waterloo, Canada, May 2–6, 2005.

[2] Morrison, J. D., *IEEE 802.11 Wireless Local Area Network Security Through Location Authentication*, Thesis from Naval Postgraduate School, Monterey, CA, September 2002.

[3] Poh, G. K. "Feasibility Study of 802.11 Distance Measurement," Thesis, Curtin University of Technology, Australia, 2005.

[4] "Wi-Fi Based Real-Time Location Tracking: Solutions and Technology," White Paper, Cisco Systems, Inc., http://cisco.com/, 2006.

[5] Ganu, S., A. S. Krishnakumar, and P. Krishnan, "Infrastructure-Based Location Estimation in WLAN Networks," *IEEE Wireless Communications and Networking Conference*, Atlanta, GA, March 21–24, 2004.

[6] Hatami, A., et al., "On RSS and TOA Based Indoor Geolocation—A Comparative Performance Evaluation," *IEEE Wireless Communication and Network Conference*, Las Vegas, NV, April 3–6, 2006.

[7] Hallberg, J., M. Nilsson, and K. Synnes, "Positioning with Bluetooth," *10th International Conference on Telecommunications ICT 2003*, Tahiti, French Polonesia, February 23–March 1, 2003.

[8] Hallberg, J., M. Nilsson, and K. Synnes, "Bluetooth Positioning," *2nd Annual Conference on Computer Science and Electrical Engineering CSEE 2002*, Lulea, Sweden, May 27–28, 2002.

[9] Feldmann, S., et al., "An Indoor Bluetooth-Based Positioning System: Concept, Implementation and Experimental Evaluation," *International Conference on Wireless Networks ICWN 2003*, Las Vegas, NV, June 23–26, 2003.

[10] Bandara, U., et al., "Design and Implementation of a Bluetooth Signal Strength Based Location Sensing System," *IEEE Radio and Wireless Conference RAWCON 2004*, Atlanta, Georgia, September 19–22, 2004.

[11] IEEE Std 802.15.4-2003, "Wireless Medium Access Control (MAC) and Physical Layer (PHY) Specifications for Low-Rate Wireless Personal Area Networks (LR-WPANs)," IEEE, October 1, 2003.

[12] Norris, M., "Location Monitoring with Low-Cost ZigBee Devices," *Embedded Control Europe Magazine*, February 2006.

[13] Aamodt, K., "CC2431 Location Engine," Texas Instruments Application Note AN042 (Rev. 1.0).

[14] Texas Instrument CC2431 Preliminary Data Sheet (Rev. 1.01), "System-on-Chip for 2.4 GHz ZigBee/IEEE 802.15.4 with Location Engine," 2006.

[15] ECMA Standard ECMA-368, "High Rate Ultra Wideband PHY and MAC Standard, 1st Edition," December 2005.

[16] Ferscha, A., W. Beer, and W. Narzt, "Location Awareness in Community Wireless LANs," *Informatik 2001*, Vienna, Austria, September 2001.

[17] Hancke, G. P., and M. G. Kuhn, "An RFID Distance Bounding Protocol," *Proceedings of IEEE/Create-Net SecureComm 2005*, Athens, Greece, September 5–9, 2005.

[18] "RFID-radar—How It Works," Trolley Scan(Pty) Ltd, http://www.rfid-radar.com/howworks.html.

CHAPTER 11

Ultrawideband (UWB)

The accuracy of time measurements for time-of-flight range estimations is a direct function of signal bandwidth. Particularly in the case of indoor location and short range systems in general, range, and position accuracy on the order of 1m demand pulse widths or rise times of several nanoseconds, and bandwidths of several hundred megahertz, with equivalent clock rates. Similar time resolution is necessary to distinguish between line-of-sight and multipath (nonline-of-sight) signals—particularly important in environments with many reflecting objects such as are encountered indoors. Better distance resolution is obtained by averaging techniques, as was described for spread spectrum systems in Chapter 3, but the ultimate distance measurement performance, considering both distance accuracy and measurement time, are derived from the signal bandwidth. Ultrawideband (UWB) communication systems are therefore especially appropriate as a platform for ranging and location.

11.1 Telecommunication Authority Regulations

Regulating the use of radio communications signals having very wide bandwidths necessitated a completely new approach to spectrum allocation. It was not feasible to assign frequency channels of such widths to specific users or for specific applications, as is common with narrowband signals, because free spectrum is simply not available. On the other hand, ultrawideband technology has the potential for important applications, among them those that concern public safety and emergency response, that are hard to accomplish with conventional wireless communication methods. The allocation solution was to overlay ultrawideband spectrum on channels that are occupied by narrowband (and what is normally considered wideband) users and to constrain power density to levels that will not cause interference. So, by constraining power density to values previously designated as limits to man-made random noise, while allocating spectrum widths on the order of many gigahertz, the total average transmitted power—density times bandwidth—is sufficient for a wide range of applications, mostly in indoor environments, that operate over distances of tens and even hundreds of meters.

11.1.1 FCC Regulations

The definition of what constitutes ultrawideband signals is clearly stated in Part 15 of the FCC regulations. A UWB transmitter is "An intentional radiator that, at any point in time, has a fractional bandwidth equal to or greater than 0.20 or has a UWB bandwidth equal to or greater than 500 MHz, regardless of the fractional bandwidth" and "The UWB bandwidth is the frequency band bounded by the points that are 10 dB below the highest radiated emission, as based on the complete transmission system including the antenna" [1]. Designating the upper boundary as f_H and the lower boundary as f_L, the fractional bandwidth equals $2(f_H - f_L)/(f_H + f_L)$. A signal spectrum with the UWB bandwidth definition showing the frequency of highest radiated emission, f_M, and center frequency, f_c, as well as a bandwidth plot complying with the fractional bandwidth requirement, are shown in Figure 11.1.

Since UWB must coexist with numerous narrowband signals that occupy its spectrum, limitation of interference is achieved by specifying power density rather than average transmitted power. Frequency bands and power densities are specified by the FCC according to categories of applications. The two categories that are intended for most commercial uses are indoor UWB systems and handheld devices under which the UWB bandwidth defined above must be contained between 3,100 and 10,600 MHz. Radiation power limits versus frequency, above 960 MHz, for these categories are shown in Figure 11.2. Power is measured at a resolution bandwidth of 1 MHz, except for two frequency ranges within the segment identified as GPS band, where the resolution bandwidth is allowed to be as low as 1 kHz—a necessary condition for measuring for compliance with the particularly low power density specified in those ranges. UWB bandwidth as defined above must be within the band from 3.1 to 10.6 GHz, where average power density is −41.3 dBm/MHz. There is also a limit on the peak level of the emissions, equal to 0 dBm equivalent isotropic radiated power (EIRP), contained within a 50-MHz bandwidth centered on the frequency, f_M, at which the highest radiated emission occurs.

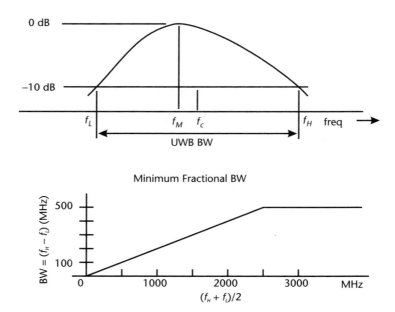

Figure 11.1 FCC definitions of UWB.

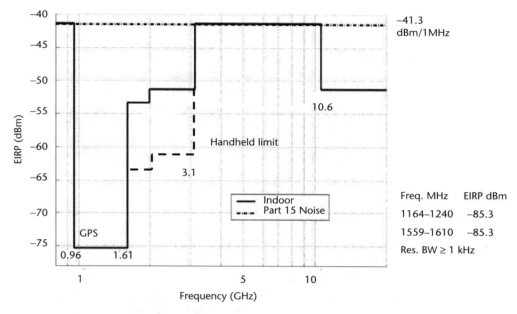

Figure 11.2   FCC emission limits for UWB indoor and handheld devices.

Radiated emissions at or below 960 MHz must not exceed the levels specified in Section 15.209 of the FCC regulations. The field strength limit at 3m for frequencies between 216 and 960 MHz is 200 microvolts/meter, measured using a CISPR (the international special committee on radio interference) quasi-peak detector.

11.1.2   UWB in the European Community

In the European Community, frequency bands for UWB are more restricted than those of the FCC although the bandwidth definition and power density levels in the transmission bands are the same. Frequency bands where the maximum average power density level of −41.3 dBm/MHz is allowed are 6 to 8.5 GHz and also 4.2 to 4.8 GHz, until the end of 2010. The bands 3.4 to 4.2 GHz, and 4.2 to 4.8 GHz without the time limitation, can be used with the −41.3 dBm/MHz upper limit provided appropriate mitigation techniques are applied. These techniques must limit the duty cycle to maximum 5 percent per second and 0.5 percent per hour, and restrict single transmission time to maximum 5 ms [2].

In addition to the average power density restrictions, the European regulations apply a maximum peak power limit measured over 50 MHz. This limit is 0 dBm within 4.2 to 4.8 GHz (until the end of 2008 and −30 dBm beyond that date) and within 6.0 to 8.5 GHz. The peak power limits are considerably reduced on other bands, where the maximum average power density level limit is well below −41.3 dBm.

The relatively low radiated power levels permissible for UWB is a limiting factor for both high-rate communications and distance measurement and location, but still a wide class of commercial applications is expected to make UWB a prominent feature of wireless devices in the years to come.

11.2 UWB Implementation

Two methods have been developed for UWB that meet the technical requirements of the regulating authorities: impulse radio (IR) and multiband OFDM (MB-OFDM). The former is the legacy method of achieving wide bandwidth, dating back to the origins of radio. It is based on creating a sequence of short pulses, modulated by pulse amplitude modulation (PAM) or pulse position modulation (PPM). MB-OFDM uses an inverse FFT to place data in spaced discrete sidebands to create a flat-topped wideband spectrum. Further widening is achieved by frequency hopping over a number of consecutive OFDM bands.

11.2.1 Impulse Radio UWB

A common form of impulse radio that is used for communication and distance measurement is time-hopping spread spectrum. It shares some characteristics with those of direct sequence spread spectrum described in Chapter 3. The wide bandwidth of IR-UWB signals is achieved by sequences of narrow pulses. We give an example of a UWB signal based on the pulse shown in Figure 11.3 and expressed as follows [3]:

$$w(t) = \left[1 - 4\pi \cdot \left(\frac{t}{\tau_m} \right)^2 \right] e^{-2\pi \left(\frac{t}{\tau_m} \right)^2} \tag{11.1}$$

The pulse width is τ_m.

When the time units of Figure 11.3 are nanoseconds and $\tau_m = 0.2877$ ns, the spectrum of the pulse is as shown in Figure 11.4, with frequency scale units GHz. The pulses are generated at baseband and conventional up conversion to RF is not required. The space on the time axis within which one pulse is present is referred to as a chip, in analogy to DSSS. Two preferred methods of pulse modulation are polarity, or bipolar, modulation and pulse position modulation. The data band-

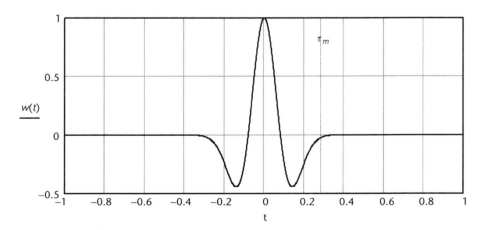

Figure 11.3 Example of UWB pulse.

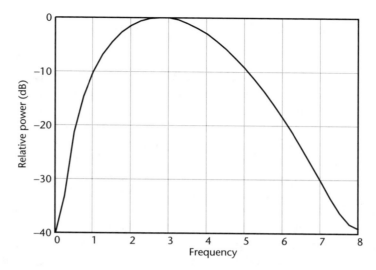

Figure 11.4   Spectrum of UWB baseband pulse.

width is much lower than the signal bandwidth, as in all forms of spread spectrum communication.

The data bandwidth is spread by a pseudorandom sequence of pulses that make up one symbol. Figure 11.5 shows an example of a time-hopping spreading sequence [3]. Each *pulse* has a width of T_P. Within each *chip* time frame, T_C, the pulse is contained in a slot T_P wide whose position is defined by a pseudorandom code sequence $c_0, c_1, \ldots c_j \ldots c_{N-1}$. c_j is an integer whose value is between 1 and N_H, the number of pulse slots per chip. $N_H = 4$ in the example of Figure 11.5. The code is defined over N chips and has a period of NT_C. In Figure 11.5, $N = 5$. A *symbol* is defined as a sequence of N chips, with period $T_S = NT_C$. Multiple UWB signals having different spreading codes can exist in the same area with a low probability of collision.

The following expression describes an unmodulated transmitted time-hopping UWB signal over unbounded time:

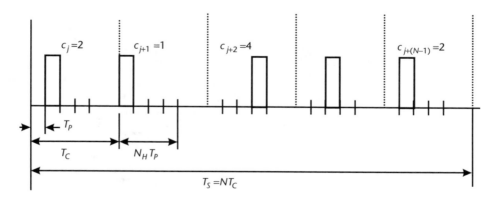

Figure 11.5   UWB time-hopping spreading sequence.

$$s(t) = \sum_{j=-\infty}^{\infty} [w(t - jT_c - c_j T_p)] \qquad (11.2)$$

where $w(t)$ is the transmitted waveform of a single pulse, T_c is the chip period and T_p is the width of the pulse slot. The code $\{c_j\}$ repeats itself every N pulses.

Note that the pulse slots covered by the hopping code may not cover the whole chip period. A time interval is left within the chip period after the last slot for channel power delay to prevent interference between multipath returns and the next pulse. This time interval can also be provided for pulse position modulation, when $N_H T_P \le 0.5 T_C$. After symbol boundaries have been determined by a synchronizing sequence in a preamble, a data level, say "one," can be defined as occurring when the coded pulse position occupies the earlier part of the chip, and "zero" when it is in the later part of the chip. Indication of the start of a data message or determination of time of arrival of a defined epoch may be facilitated by starting a transmitted packet with a preamble containing a string of pulses with a given predetermined code, organized as shown in Figure 11.5, followed by a start of frame deliminator (SFD)—a known sequence of data symbols whose last arriving symbol boundary marks the beginning of a data message.

There are two basic architectures of IR-UWB receivers, coherent and noncoherent. A simplified architecture of a correlation receiver is shown in Figure 11.6. The template generator produces a replica of the known pulse sequence in a symbol. This reference signal is multiplied by the received signal. The product is integrated over the symbol period and applied to the peak detector, whose output controls the timing, or phase difference, between the reference and the received signal. Each symbol period the phase is adjusted by a small amount, no greater than the chip period, until the integrator output exceeds a given threshold. Subsequent smaller adjustments of the reference timing will produce a correlation peak, indicating that the symbol timing in the received signal is synchronized with the receiver symbol clock.

Once symbol synchronization has been achieved, the symbol bits can be detected. Data is imposed on the hopping sequence by using the polarity of the pulses in a symbol sequence to indicate a bit level of "zero" or "one." This method is called bipolar modulation. A second way of distinguishing the bits is to shift the

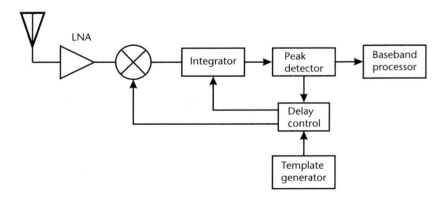

Figure 11.6 Coherent IR-UWB receiver front end.

position of the pulse burst within the symbol, in relation to the symbol boundaries, as previously described. This is pulse position modulation (PPM). Data detection can be performed by using two reference templates in Figure 11.6, together with two parallel chains of multiplier, integrator, and peak detector. The data bit of each symbol is determined by the peak detector with the highest output.

As stated previously, a start frame deliminator (SFD) is a known data sequence at the end of the synchronization period of the preamble. Once symbol synchronization has been achieved, the baseband processor in Figure 11.6 searches for the SFD. The end of the last symbol in the SFD is the beginning of the message in the transmitted packet, and also indicates the epoch whose time of occurrence is the time of arrival of the signal (TOA) that is used for distance measurement and location.

In order for the correlation process described above to be effective, the pulse shape created in the template generator of Figure 11.6 must be very similar to that of the received pulse. Therefore, the frequency response of transmitting and receiving antennas, as well as the impulse response of the transmission path, have to be taken into account. Also, multipath propagation can cause the receiver to synchronize on a reflected version of the transmitted signal, instead of the desired line-of-sight (LOS) signal, resulting in a late TOA that will cause a distance estimation to be greater than the true distance between the two terminals. A ranging receiver having a single correlator would have to continuously repeat the symbol synchronization process until the path with the earliest symbol boundary time is found. As a faster alternative, multiple correlators operating in parallel are used, each fed with a different delay of the reference sequence. Such an arrangement shortens the signal acquisition time and also indicates the shortest path.

Much of the complexity of the UWB correlation receiver is avoided by using a noncoherent architecture, shown in Figure 11.7. Instead of the correlator and internally generated reference, the received signal is multiplied by itself in a squaring block. The squarer output is integrated and then applied to an energy detector, whose function is similar to that of the peak detector in the correlation receiver. The feedback function that controlled the relative phase of the reference signal in Figure 11.6 operates in the noncoherent receiver of Figure 11.7 to adjust the start of integration. The control input to the integrator is a sequence of square pulses, corresponding to the time hopping pulses that are transmitted. When the control

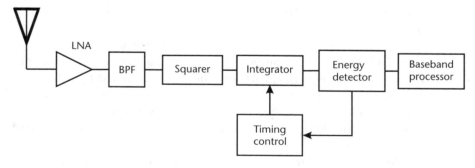

Figure 11.7   Noncoherent IR-UWB receiver front end.

pulses are high, signal plus noise power is accumulated, and integration is suspended when those pulses are low. At the end of a symbol period, the output level of the integrator is transferred to the energy detector and the integrator is reset, ready to start a new period of signal power accumulation.

The highest output of the integrator will most likely occur when each of the square pulses in the sequence from the timing control block encompasses, that is, is in time with, the received pulses, because the energy of the signal plus noise during the pulse duration is greater than that of noise alone. The energy detector commands the timing control block to adjust the phase of the square pulses to the integrator control input in relation to the incoming signal, until an energy threshold is exceeded. This threshold is chosen as a compromise between false detects, when it is set too low, and synchronization misses, when it is set too high.

As in the case of the correlation receiver, multiple paths can cause synchronizing on a reflected signal. If the UWB duty cycle is very low and the period between pulses is greater that the duration of all significant echoes, the integration period corresponding to the start of each pulse can be set to include the direct pulse and all its major reflections.

While the noncoherent energy detection receiver is simpler than the coherent receiver, it does have disadvantages. It requires a higher signal-to-noise ratio for a given error rate because the noise and interference power is integrated together with the signal and the squaring is apt to increase the response to noise. The bandpass filter before the squarer must have minimum bandwidth possible while retaining the transmitted pulse shape. In a correlator, the signal-to-noise ratio is the ratio of the energy per symbol period and the noise density, and is not a function of the input bandwidth. Another characteristic of the energy detection receiver is that it cannot demodulate bipolar pulses. It can be used for PPM, or on-off keying (OOK), where logic levels are discerned by existence or nonexistence of a pulse in the chip period.

11.2.2 OFDM

Another method for creating an ultrawideband signal that meets FCC requirements and has features attractive to short-range distance measurement and location is orthogonal frequency division multiplex (OFDM). OFDM is particularly efficient for capturing the total energy in multipath channels. It has high spectral efficiency due to its spectrum shape, which is almost completely flat over its bandwidth between extremities of 10 dB down. The properties of OFDM make it possible to exclude from the spectrum explicit frequency ranges—an advantage since UWB frequency coverage is not identical around the world.

An OFDM signal is composed of individual subcarrier frequencies on both sides of a center frequency, each of which is modulated by a bit or small subset of bits in a data sequence. The length of the data sequence is an OFDM symbol period. Modulation is MPSK (M-ary phase shift keying) or QAM (quadrature amplitude modulation), so each subcarrier has constant amplitude and phase during a symbol. The data rate in each subcarrier is the total data rate times the number of bits in a subset of bits that modulate each subcarrier divided by the number of bits in the sequence. Adjacent subcarriers are orthogonal and do not interfere. The fact that the OFDM symbol period is much larger than the period of the bits in the source data stream is responsible for the relative immunity of an OFDM symbol to multipath and intersymbol interference.

The OFDM signal is described as follows. The bits in a sequence are translated to complex representations of the subcarriers, arranged in vector form $(C_0, C_1 \ldots C_{N-1})^T$ (the superscript T indicates transpose). If binary phase shift keying (BPSK) modulation is to be used, for example, the phase of each C_i may be either $+\pi/2$ or $-\pi/2$ representing logic level 0 or 1. With two bits per symbol in QPSK (quadrature phase shift keying), the phase of $C_i \in \{0, \pi/2, \pi, -\pi/2\}$. Thus, during one sequence of bits, the phase and relative amplitude of each subcarrier is determined by C_i. Each sequence vector undergoes a discrete inverse Fourier transform [implemented by an inverse fast Fourier transform (IFFT)] with samples given as:

$$x(n) = \sum_{k=0}^{N-1} C_k e^{j2\pi \cdot \frac{k}{N} n} \tag{11.3}$$

where N is the total number of subcarriers. Equation (11.3) is expressed in terms of frequency and time parameters as:

$$s(t) = \sum_{m=-(N/2)+1}^{N/2} C_k e^{j2\pi \cdot m \cdot f_0 \cdot t} \tag{11.4}$$

where f_0 is the frequency separation between subcarriers, $f_0 = f_s/N$, and f_s is the sampling frequency. After an upconversion to carrier frequency f_c, the resulting OFDM RF signal is:

$$s_{rf}(t) = \mathrm{Re}\{s(t) e^{j2\pi \cdot f_c \cdot t}\} \tag{11.5}$$

Figure 11.8 is a basic block diagram of an OFDM transmitter-receiver. In the transmitter, data is converted directly to amplitude and phase of the subcarriers and then converted by IFFT to the time domain where it modulates the center frequency carrier. A reverse process takes place in the receiver. A guard band is included in the time domain signal for each sequence to prevent intersymbol

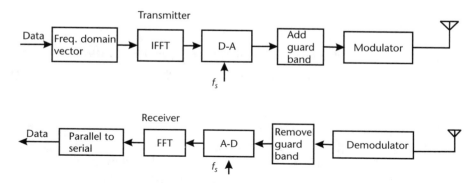

Figure 11.8 Basic block diagram of OFDM transmitter-receiver.

interference from multipath signals from a previous sequence. A simulated OFDM spectrum with 64 subcarriers is shown in Figure 11.9. Twelve subcarriers on both upper and lower extremes have zero amplitude and serve as guard bands in the frequency domain. Sampling frequency is 1 Hz and the spectrum is shown before upconversion. Note that the spectrum is flat topped, and since the FCC regulations specify a limit on the power density, the OFDM spectrum insures a possibility of achieving maximum average power for a given spectral width.

Additional spreading of the basic OFDM signal for UWB may be attained by frequency hopping. The system in which a UWB signal is created by OFDM and then consecutive packets are displaced in frequency over adjacent bands is called multiband OFDM (MB-OFDM). The process is illustrated in Figure 11.10.

Coherent synchronization of OFDM is a critical aspect of reception since coherent demodulation is required to maintain the orthogonality of the subcarriers. High-resolution TOA can be measured and used for ranging and location. Both the impulse radio and OFDM methods of creating UWB signals make possible high resolution TOA measurements and facilitate distinguishing between line-of-sight signals and multipath. Positioning for location using TOA or TDOA is carried out by methods described in Chapter 7. Ranging is done by the time transfer method of Chapter 4 and an example was given for ECMA-368, based on MB-OFDM, in Chapter 10.

11.3   IEEE 802.15.4a

IEEE 802.15.4a is an example of a standard that was conceived to provide precision ranging based on an impulse radio UWB platform. It describes an alternate physical layer to the IEEE 802.15.4 standard, which defines a physical layer and MAC functions for low-rate personal area networks, commonly known as ZigBee. The precision ranging capability in IEEE 802.15.4a is intended to satisfy industrial and consumer requirements for WPAN communications having a distance measuring accuracy of 1m or better, and with improved communication range, robustness and mobility over the original DSSS-based 802.15.4. The standard aims to meet world wide regulatory requirements to give it international relevance. Technical features of the standard were conceived to provide low complexity, cost, and power consumption and to support coexistence among sensor networks, controllers, and peripheral devices in colocated systems.

11.3.1   Physical Layer Characteristics and Synchronization

802.15.4a describes two different physical layer technologies: chirp and UWB [4]. Fourteen channels for the chirp signals, each 5 MHz wide, are defined by center frequencies that range between 2,412 and 2,484 MHz. The chirp solution in 802.15.4a does not support ranging, although proposed ranging for this technology is discussed in [5]. The UWB technology in the specification meets the FCC specification for UWB as well as that which is being developed in the European community. Fifteen frequency channels are allotted with center frequencies between 3,494.4 and 9,484.8 MHz, not including frequencies between approximately 4.8 and

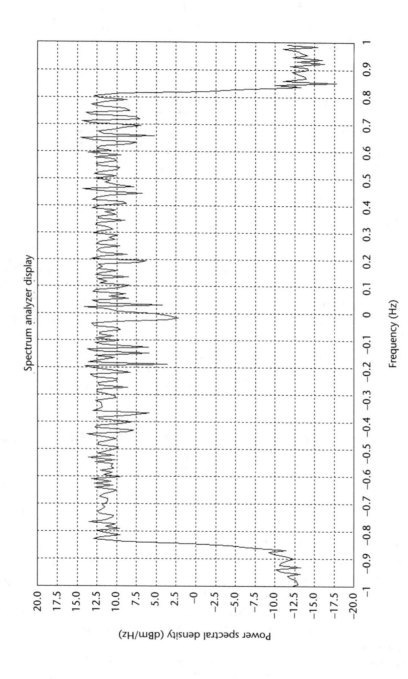

Figure 11.9   Simulated spectrum of OFDM signal.

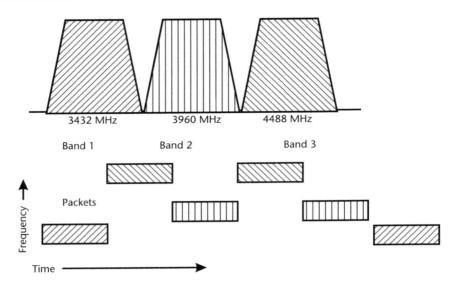

Figure 11.10 Multiband OFDM band hopping.

6 GHz in order to avoid interference with the Unlicensed National Information Infrastructure (U-NII) Devices band which is used for WLAN under the IEEE 802.11a specification. Eleven channels have bandwidths of 499.2 MHz, two channels have bandwidths of 1,331.2 MHz, and two other channels have bandwidths of 1,081.6 MHz and 1,354.97 MHz each. The channels with bandwidths greater than 1 GHz can provide extra high distance measurement resolution. An additional specified channel has a center frequency of 499.2 MHz with a bandwidth of 499.2 MHz. This lower frequency UWB channel is intended for wall and ground penetration imaging applications.

The basic element of the IR-UWB technology is the individual pulse. 802.15.4a defines a reference root raised cosine pulse, drawn in Figure 11.11 and expressed as

$$r(t) = \frac{4\beta}{\pi\sqrt{T_c}} \frac{\cos\left[(1+\beta)\pi\frac{t}{T_c}\right] + \frac{(1-\beta)\pi}{4\beta}\text{sinc}[(1-\beta)\pi t/T_c]}{1 - (4\beta t/T_c)^2} \tag{11.6}$$

where $\text{sinc}(x) = [\sin(x)]/x$ and roll-off factor $\beta = 0.6$. Parameter T_c is nominally the pulse width, which determines the bandwidth. It equals 2 ns for a bandwidth of 499.2 MHz and 0.75 ns for 1,331.2 MHz, 0.92 ns for 1,081.6 MHz and 0.74 ns for 1,354.97-MHz bandwidths [4]. The form of $r(t)$ in (11.6) was chosen to reduce intersymbol interference (ISI) in receivers having a root raised cosine type input filter. Reduced ISI improves a resolution of pulse arrival time and recognition of multipath echoes.

The cross correlation of the actual pulse used in an 802.15.4a compliant system with the reference pulse must be at least 80% of perfect correlation with sidelobes not exceeding 30% [4]. This pulse is upconverted by multiplying with a carrier wave to any of the UWB channels. Figure 11.12 is the one-sided baseband spectrum of the pulse in which $T_c = 2$ ns. Its double-sided width at −10 dBm exceeds the

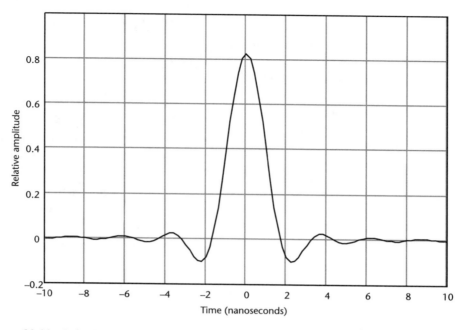

Figure 11.11   Reference UWB pulse in 802.15.4a.

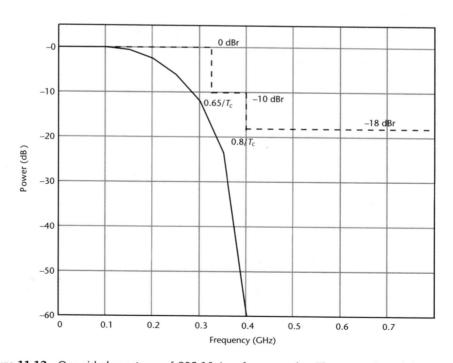

Figure 11.12   One-sided spectrum of 802.15.4a reference pulse. The spectral mask for $T_c = 2$ ns is indicated by a dashed line.

minimum FCC regulation requirement of 500 MHz while remaining within the transmit spectral mask specified in IEEE 802.15.4a, shown as a dashed line in the diagram. It should be remembered that the actual shape of a received pulse and signal spectrum depend on the characteristics of the transmitting and receiving antennas, and the impulse response of the transmission path.

Pulses like the one of Figure 11.11 are transmitted in bursts. The polarity of the pulses in a burst together with the position of the burst within a symbol period convey 2 bits per symbol, and the data rate is set by the number of pulses in a burst. Bit rates may range from approximately 0.1 to 27 Mbps, while support for 0.85 Mbps is mandatory [4].

It is instructive to know the best possible range accuracy that can be obtained from a TOA measurement. The Cramer-Rao inequality gives a lower bound for the standard deviation of the propagation time estimate τ. For a single path channel with additive white Gaussian noise (AWGN), it is expressed as

$$\sigma_\tau \geq \frac{1}{2\pi \cdot \beta \cdot \sqrt{2 \cdot (SNR)}} \tag{11.7}$$

where β is the effective signal bandwidth and SNR is the signal-to-noise ratio [6]. Using as an example $\beta = 500$ MHz, and $SNR = 10$, the lower bound is $\sigma_\tau = 0.071$ ns, equivalent to a distance of 2 cm. Errors that cause the actual TOA measurements to exceed this value are due to multipath propagation, nonline-of-sight (NLOS) reception and multiuser interference.

The synchronization preamble of an 802.15.4a packet is made up of a synchronization field (SYNC) and a start of frame (SFD) delimiter. The SYNC may contain 16, 64, 1,024, or 4,096 identical repeating symbols. The choice of SYNC length depends on the channel delay profile (impulse response) and signal-to-noise ratio. Large synchronization fields are preferred for noncoherent receivers to allow additional time for signal acquisition and frame synchronization. A SYNC symbol is made up of a sequence of 31 pulses each of which may have one of three states: plus or minus polarity, or zero value. These individual pulses are separated by idle periods whose length is a function of channel number and chosen pulse repetition frequency. The symbol used in the SYNC is chosen from the set of such ternary sequence codes that are shown in Table 11.1. The S1 sequence is drawn in Figure 11.13. A "−" in the sequence in Table 11.1 is shown as a pulse with amplitude

Table 11.1 802.15.4a Preamble Ternary Sequence Codes

Preamble Symbol	Symbol Pulse Sequence
S_1	−0000+0−0+++0+−000+−+++00−+0−00
S_2	0+0+−0+0+000−++0−+−−−00+00++000
S_3	−+0++000−+−++00++0+00−0000−0+0−
S_4	0000+−00−00−++++0+−+000+0−0++0−
S_5	−0+−00++++−+000−+0+++0−0+0000−++
S_6	++00+00−−−+−0++−000+0+0−+0+0000
S_7	+0000+−0+0+00+000+0++−−−0−+00−+
S_8	0+00−0−0++0000−+00−+0++−++0+00

Source: [4].

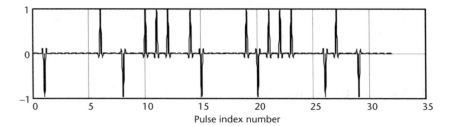

Figure 11.13   Basic ternary sequence S1. Range of pulse index numbers is 1 to 31.

−1, and a "+" in the table is a +1 amplitude pulse in the figure. IEEE 802.15.4a specifies two allowed sequences from the table for each channel, chosen for their very low cross correlation. This means that two adjacent PANs can operate with no interference between them, effectively doubling the possible number of independent channels to 32—twice the number of UWB frequency channels.

Each of the ternary sequences has perfect autocorrelation, that is, it has a peak value of unity when the receiver generated replica is perfectly lined up with the received sequence and no sidelobes when the replica is displaced in time by more than one pulse width. Perfect correlation is retained when the input sequence is squared, as it is in a noncoherent receiver. The receiver synchronizes its local reference sequence to the SYNC symbols of the received signal. It must then identify the epoch in the packet where the time of arrival measurement is made—at the end of the synchronizing preamble and the start of the physical layer (PHY) header. It does this by searching for and identifying the start of frame deliminator (SFD). The SFD may contain 8 or 64 symbols. The default short sequence has 8 symbols and the long, optional 64-symbol SFD is used with the low data rate of 110 kbps. Each SFD symbol is the same as the basis symbol of the SYNC, one of which is depicted in Figure 11.13. This symbol can be expressed mathematically as [6]:

$$w_i(t) = \sum_{j=0}^{L-1} S_i(j) \cdot \phi(t - jT_{pri}) \qquad (11.8)$$

where $L = 31$ is the length of the basis sequence code, $\varphi(t)$ is the waveform of an individual pulse, and T_{pri} is the pulse repetition interval—the period between adjacent pulses that includes the added idle period that spreads out the pulses in the symbol. Then the SFD can be written as:

$$Z_i(t) = \sum_{m=0}^{L_{SFD}-1} M(m) \cdot w_i(t - mT_{sym}) \qquad (11.9)$$

where T_{sym} is the length of the basis SYNC symbol and L_{SFD} is the number of symbols in the SFD—8 or 64. $M(m)$ is a component of a vector of length L_{SFD} that can equal 0, +1, or −1.

For example, the vector **M** of the short sequence is [0 1 0 −1 1 0 0 −1]. Then, the SFD sequence is {0, $+S_i$, 0, $−S_i$, $+S_i$, 0, 0, $−S_i$}, where S_i is the symbol used in the preamble and the zeros represent a symbol time of no transmission. After the receiver demodulates, or despreads, the SFD by correlating with the known sequence signal $Z(t)$, it can recognize precisely the end of the last symbol in the synchronization preamble, whose time of occurrence is the TOA used in 802.15.4a ranging.

11.3.2 Ranging Protocol

The measurement of time of arrival of an epoch of the received packet is used in a two-way protocol to find the distance between two 802.15.4a terminals. The protocol is based on a time transfer routine such as was described in Chapter 4. Considering the duration of time transfer between two terminals, the fact that their clock rates are not synchronized is a source of inaccuracy in the range determination. The protocol incorporates measures to neutralize this source of error.

The ranging operation consists of several back and forth packet exchanges, shown schematically in Figure 11.14. We will call the initiator of the measurement side A and the responder, or target, side B. The basic ranging exchange is as follows. Side A sends a ranging packet to side B, noting a counter value, t_{1A}, at the instant that the SFD of its packet leaves the antenna of side A. Side B receives the packet and notes the time t_{1B} of reception at the end of the SFD. Side B then returns an acknowledge packet, noting the time t_{2B} of the end of SFD epoch as it leaves its antenna. At this point, when two counter values, t_{1B} and t_{2B}, have been recorded, side B forms a timestamp report, whose contents are described below. Side B's acknowledgment (Message 2) is received at side A, which records its counter value, t_{2A}, when the SFD has been received. Side A then makes a timestamp report containing t_{1A} and t_{2A}. The timestamps are used by a range estimating function, which may be the originator or any other designated terminal, to calculate the distance. If side A does the ranging, side B sends it its timestamp in a normal message packet. Side A uses the timestamp that it receives from side B, together

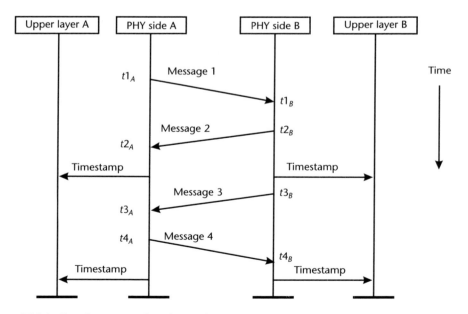

Figure 11.14 Ranging protocol packet exchanges.

with its recorded counter values t_{1A} and t_{2A}, to calculate the range estimate. The basic uncorrected distance estimate is:

$$d = \frac{c}{2} \cdot [(t2_A - t1_A) - (t2_B - t1_B)] \qquad (11.10)$$

where c is the speed of light.

In addition to start (transmission) and stop (reception) counter values, parameters in the timestamp facilitate correction of the distance estimation to account for clock drift between the two sides, and to assess the accuracy of the measurement. The timestamp report contains five parameters in a total of 16 octets. The contents of the timestamp are shown in Table 11.2.

The resolution of the start and stop time parameters is the mandatory chip time of 2 ns, divided by 128, or approximately 16 ps. The maximum recordable value is 67 ms. The start time value corresponds to a range counter reading at the time a message epoch is at the antenna at the beginning of a ranging message exchange whereas the stop time value indicates the reading at the end of the exchange. The differences of the readings are used as indicated in (11.10). Tracking interval and tracking offset values correct for clock drift when the devices support the feature of crystal characterization. Tracking offset counts time units that the receiver advances or retards its clock in order to maintain symbol synchronization during preamble reception. The tracking interval is the number of time units over which tracking offset is measured. The expression for the corrected measurement time interval is:

Measurement_interval = (stop_time − start_time) × (1 + tracking_offset/tracking_interval)

$$(11.11)$$

We have seen that the range calculation (11.10) uses the difference of the stop time and start time values when calculating range, but the fact that they are sent individually in the timestamp packet allows counter value t_{2A} in Figure 11.14 to be used in a network that estimates location by the TDOA method. In this case a network infrastructure function would use the differences between $t2_A$ values, reported after range transactions between several side A terminals with the side B target as time difference of arrival parameters to find location as described in Chapter 7. The counter clocks for the participating side A terminals must be synchronized for this operation.

The fifth parameter, figure of merit (FOM), in the timestamp report gives an indication of the accuracy of the arrival time estimate of the epoch at the end of

Table 11.2   802.15.4a Timestamp

Parameter	Number of Bits
Start time	32
Stop time	32
Tracking interval	32
Tracking offset (signed)	24 (4 not used)
Figure of merit (FOM)	8

the SFR, or beginning of the physical layer header. It is a measure of the precision of discerning the leading edge of the received pulse that specifies the time of arrival. The FOM octet has three subfields: confidence level, confidence interval, and a scaling factor. The confidence level—one of seven values between 20% and 99%—is a measure of the confidence that the measured time is within the confidence interval. The value of 2 bits in the confidence interval subfield points to a confidence interval of 100 ps, 300 ps, 1 ns, or 3 ns, which is multiplied by the scaling factor. Two bits are used for the scaling factor, which may be 1/2, 1, 2, or 4. An additional bit in the FOM field is used to indicate that the ranging counter start value cannot be used at all.

The figure of merit in the timestamp reflects several aspects of the TOA measurement: signal-to-noise ratio of the link, the length of time after detection of the preamble that is used for searching for the earliest multipath pulse sequence, and the self-calibrating capability of the device [7].

As an optional measure to reduce the effect of time base drift between the two terminals when crystal offset correction is not available, double-sided two-way ranging can be employed as shown in Figure 11.14 (see the discussion on back-to-back ranging in Section 4.3.1). In this case, after sending its acknowledgment (message 2) to side A, side B sends a ranging packet (message 3) to A, noting the epoch transmission time t_{3B}. Side A records the counter value of receipt of the packet as t_{3A}. Subsequently, side A sends an acknowledgment (message 4) to side B. A range estimate that largely cancels out the relative clock drift is:

$$d = \frac{c}{4} \cdot [(t2_A - t1_A) - (t2_B - t1_B) + (t4_B - t3_B) - (t4_A - t3_A)] \quad (11.12)$$

Figure 11.14 shows that for symmetrical two-way ranging, message 3 from side B to side A follows message 2 in the same direction. Message 3 is thus superfluous, and only three messages are needed for symmetrical two-way ranging—one more than required for two-way ranging where the crystal offset correction facility needs to be available for high precision. In this case, $t_{3B} = t_{2B}$ and $t_{3A} = t_{2A}$. Making the substitutions, (11.12) can now be rewritten:

$$d = \frac{c}{4} \cdot [2(t2_A - t2_B) + (t1_B - t1_A) + (t4_B - t4_A)] \quad (11.13)$$

It is important to realize that the start time and stop time measurements (Table 11.2) refer to the instants that message epochs leave or arrive at the device's antenna. 802.15.4a provides for a self-calibrating capability where the device can account for the time between the arrival of a signal epoch at the antenna and the time of reading the ranging counter, and similarly for a transmitted packet, the time difference between the counter reading and the departure of the packet from the antenna. Thus, the calibration function allows compensation for the signal delay in the receiver and transmitter front end circuitry. These delays could be significant considering the desired time of flight precision of the order of a nanosecond.

802.15.4a includes specifications for an optional private ranging protocol, for use in networks that require security against eavesdropping or obstruction of operation [6]. The ranging operation commences when side A sends a range authentication packet whose purpose is to allow the target, side B, to authenticate the originating device. It also includes encrypted identifiers of two special 127-length ternary sequences to use in the preambles of the subsequent ranging initiation and acknowledgment. The substituted preambles reduce the probability of eavesdropping on the ranging exchange and initiating ranging by an unauthorized device. Encryption of the timestamp reports prevents outsiders from learning range information.

11.4   Dealing with Multipath and Nonline of Sight

While UWB systems make accurate and precise distance estimations when there is a dominant line-of-sight (LOS) path to the target, multiple reflections and physical obstructions between the communicating terminals can cause a range estimate to deviate considerably from its true value. When there are multiple reflections, the time of the earliest arriving one will be the best measurement value to use for the range calculation. A UWB ranging device must therefore be designed to search for and acquire multiple time-delayed versions of the transmitted sequence and to use the earliest arriving one for distance or location determination. It is also desirable for the device to recognize instances where the earliest detected path is not the line of sight.

11.4.1   Multipath

We can represent the creation of the multiple returns over multiple paths as follows. The impulse response of the channel can be approximated as a succession of impulses, each occurring at a specific time delay and multiplied by a complex amplitude factor [8]:

$$h(t) = \sum_{i=1}^{L} \alpha_i \cdot \delta(t - \tau_i), \ \tau_1 < \tau_2 \ldots \tau_L \qquad (11.14)$$

where L is the number of paths, α_i is the amplitude and τ_1 is the direct path delay and α_1 is its amplitude. Let the transmitted pulse sequence be $s(t)$, for example, as expressed in (11.2). Then the received signal is:

$$r(t) = s(t) \ * \ h(t) + n(t)$$

$$r(t) = s(t) \ * \ \left[\sum_{i=1}^{L} \alpha_i \cdot \delta(t - \tau_i) \right] + n(t) \qquad (11.15)$$

$$r(t) = \sum_{i=1}^{L} \alpha_i \cdot s(t) \ * \ \delta(t - \tau_i) + n(t)$$

$$r(t) = \sum_{i=1}^{L} \alpha_i \cdot s(t - \tau_i) + n(t), \ \tau_1 < \tau_2 \ldots \tau_L$$

where $n(t)$ is the noise and interference and * denotes convolution. The last form of (11.15) utilizes the sampling nature of convolution with an impulse [9]:

$$g(u) * \delta(u - u_0) = g(u - u_0) \tag{11.16}$$

In the time-hopping UWB system discussed in Section 11.2, the receiver looks for a sequence with known code times between each pulse in the sequence. In IEEE 802.15.4a the receiver searches for a sequence of ternary pulses. It is evident from (11.15) that there are L such sequences for each transmitted sequence, and the receiver may lock on to any one of them. However, the receiver has to identify the direct path, whose delay τ_1 is the desired time of arrival. The TOA estimate process can be simply defined by two steps [10]:

1. Lock on a known sequence of the incoming signal.
2. Search for other identical sequences with smaller delays, until the sequence with the smallest delay (τ_1) has been located.

The first step can be relatively time consuming, particularly when the time between pulses is large compared to the pulse duration (low duty cycle). The preliminary search is preferably performed by a number of correlators or energy sensing detectors operating in parallel. The period between pulses is divided up into separate windows, each tested by one of the detectors. A comparison block decides which window, if any, contains the pulse or a complete sequence. This coarse acquisition is followed by a fine adjustment of the clock to the start of the pulse or sequence. Synchronization algorithms are described in [11–14].

The second step proceeds to search for sequence arrival times earlier than the first. An iterative search process is described in [15].

11.4.2 Nonline of Sight

It is likely, particularly in indoor environments, that the earliest TOA discovered by the search procedure will not be over the direct path between the two terminals. The amplitude over the line-of-sight path may be too low to be detected in the background noise, or the path may be completely obstructed. In this case, the estimated range will be too high. Even when averaged over multiple received packets, the final estimate will be too large and the data is considered to have a biased mean.

It is possible to distinguish between an LOS estimate and NLOS estimate of the TOA. According to [10]. the variance of a NLOS TOA result is larger than that of LOS. The variance of an estimate is found from the statistics of measurement data. If N measurements are made of time of arrival, $\tau_1, \ldots \tau_N$, then the sample mean is

$$\hat{\tau}_{TDA} = \frac{1}{N} \cdot \sum_{i=1}^{N} \tau_i \tag{11.17}$$

and the sample variance is

$$\hat{\sigma}_{TDA}^2 = \frac{1}{N-1} \sum_{i=1}^{N} (\tau_i - \hat{\tau})^2 \tag{11.18}$$

In (11.18) $\hat{\sigma}_{TDA}^2$ is an unbiased estimate, as is the mean in (11.17), that is, the expectation (the average value) of the samples over a large number of trials of the estimate equals the true value for all values of the time of arrival.

The variance of the line-of-sight estimate is found from knowing the system noise statistics [10]. By comparing σ_{LOS}^2, calculated from known noise statistics, and $\hat{\sigma}_{TOA}^2$, based on empirical data, some indication of the goodness of the TOA estimate is ascertained even if only qualitatively. In a location system where there are a number of fixed stations whose estimated ranges to a target participate in the positioning calculations, the estimates that are found to be NLOS can be disregarded, or at least given a diminished weight in the calculations in order to improve the final estimate of location.

The least squares technique of estimation location is optimal for unbiased distance measurements between at least three fixed stations and a target, in the case of unambiguous two-dimensional positioning. Let the known coordinates of the fixed stations be expressed in vector form as

$$\mathbf{P_i} = \begin{bmatrix} x_i \\ y_i \end{bmatrix} \tag{11.19}$$

and the target as

$$\mathbf{P} = \begin{bmatrix} x \\ y \end{bmatrix} \tag{11.20}$$

The estimate of $\mathbf{P}$, $\hat{\mathbf{P}}$, is [10]:

$$\hat{\mathbf{P}} = \arg \min_{\mathbf{P}} \left\{ \sum_{i=1}^{N} \alpha_i \left(r_i - \|\mathbf{P} - \mathbf{P_i}\| \right)^2 \right\} \tag{11.21}$$

where α_i is a weighting factor for distance measurement estimate r_i between $\mathbf{P_i}$ and the target and N is the number of fixed stations. The notation $\|P - P_i\|$ expresses the distance between the target and base station position vectors represented by $\mathbf{P}$ and $\mathbf{P_i}$ and which is also written as

$$\|P - P_i\| = \sqrt{(x_i - x)^2 + (y_i - y)^2} \tag{11.22}$$

The weighting factor α_i can be proportional to the signal-to-noise ratio for path i at the measuring terminal with k a convenient constant:

$$\alpha_i = k \cdot SNR_i \tag{11.23}$$

Expression (11.21) is interpreted as follows. The value of **P** within the braces { } is unknown. An assumed value of the vector **P**, the coordinates of the target, is chosen and the summation is performed. Next, a different value of **P** is used in the equation, which results in a new value of the summation. When all possible coordinates of **P** are tried, for one them the summation will be minimum. The value of **P** that gives this minimum is estimated to be the position of the target. There are infinitely many values of **P**, so obviously not all are tried. The test values of **P** are those lying on a grid with the required resolution of the target position.

As mentioned above, NLOS estimations of r_i are biased and their use in (11.21) will spoil the estimate of **P**. Instead of disregarding the distance measurements that are found to be NLOS, [10] presents an alternate rule to (11.21) for obtaining a most likely estimate. This rule is based on an IEEE channel model for a high rate UWB personal area network that shows multipath arrival time as following a Poisson distribution. With such an assumption, statistical properties of the LOS TOA can be derived from the times and variances of signal arrival times over the NLOS paths.

11.5 Conclusions

UWB communication techniques are best able to provide high accuracy ranging using time-of-flight (TOF) methods. Their wide bandwidth translates to high resolution timing of pulse arrivals, and the use of narrow pulses lets them distinguish between multipath returns. However, telecommunications regulations limit the power density and subsequently the average power of transmitted UWB signals. The low power of transmitted UWB systems makes them applicable particularly to indoor short range requirements where multipath is especially severe. The use of personal area network (PAN) communications for low and high rate communications has instigated applications for distance measurement and location as a supplement to data communications. The short distances involved require distance measurement accuracy of better than 1m, which is within the capability of the UWB system.

Along with the requirement for high ranging and positioning accuracy, UWB systems must be low cost and have low power consumption. The IEEE 802.15.4a specification described in this section was conceived to meet these requirements and to answer the regulatory conditions worldwide. It is to be expected that UWB ranging and location systems will make their place for short range indoor applications in parallel to the place GPS has and will continue to occupy for general geolocation purposes.

References

[1]   Federal Communications Commission, Code of Federal Regulations, Title 47, Part 15, Section 15.503.

[2]   Commission of the European Communities, "Commission Decision on Allowing the Use of the Radio Spectrum for Equipment Using Ultra-Wideband Technology in a Harmonised Manner in the Community," Brussels, February 21, 2007.

[3]   Win, M. Z., and R. A. Scholtz, "Ultra-Wide Bandwidth Time-Hopping Spread-Spectrum Impulse Radio for Wireless Multiple-Access Communications," *IEEE Transactions on Communications*, Vol. 48, No. 4, April 2000.

[4]   IEEE P802.15.4a^{TM}/D7, "Draft Amendment to IEEE Standard for Information Technology—Telecommunications and Information Exchange Between Systems—Part 15.4: Wireless Medium Access Control (MAC) and Physical Layer (PHY) Specifications for Low-Rate Wireless Personal Area Networks (LR-WPANs): Amendment to Add Alternate PHY," January 2007.

[5]   Lampe, R. R. Hach, and L. Menzer, "Chirp Spread Spectrum (CSS) PHY Presentation for 802.15.4a" *IEEE P802.15 Working Group for Wireless Personal Area Networks (WPAN)*, Doc. IEEE P.802-15-002-00-004a, 2004.

[6]   Sahinoglu and Z., S. Gezici, "Ranging in the IEEE 802.15.4a Standard," *IEEE Annual Wireless and Microwave Technology Conference, 2006 (WAMICON 06)*, Clearwater Beach, FL, December 2006, pp. 1–5.

[7]   Brethour, V., "Ranging with Draft 2," *IEEE P802.15 Working Group for Wireless Personal Area Networks (WPAN)*, Doc.IEEE 15-06-0242-00-004a, May 15, 2006.

[8]   Rabbachin, A., et al., "Non-Coherent Energy Collection Approach for TOA Estimation in UWB Systems," *14th IST Mobile & Wireless Communications Summit*, Dresden, June 19–23, 2005.

[9]   Carlson, A. B., *Communication Systems: An Introduction to Signals and Noise in Electrical Communication*, New York: McGraw-Hill, 1968, p. 46.

[10]   Sahinoglu and Z., S. Gezici, "UWB Geolocation Techniques for IEEE 802.15.4a Personal Area Networks," *Mitsubishi Electric Research Laboratories TR-2004-110*, August 2004.

[11]   Nekoogar, F., F. Dowla, and A. Spiridon, "Integration Window Position Estimation in TR Receivers," *WirelessCom2005*, Maui, HI, June 13–16, 2005.

[12]   Cheng, X., and D. Anh, "A Synchronization Technique for Ultrawideband Systems Using IEEE Channel Models," *IEEE CCECE/CCGEI*, Saskatoon, May 2005.

[13]   Carbonelli, C., and U. Mengali, "Synchronization of Energy Capture Receivers for UWB Applications," *13th European Signal Processing Conference*, Antalya, Turkey, September 5–8, 2005.

[14]   Djapic, R., et al., "Blind Synchronization in Asynchronous UWB Networks Based on the Transmit-Reference Scheme," *EURASIP Journal on Wireless Communications and Networking*, Vol. 2006, Article ID 37952, pp. 1–14.

[15]   Lee, J. Y., and R. A. Scholtz, "Ranging in a Dense Multipath Environment Using an UWB Radio Link," *IEEE Journal on Selected Areas in Communications*, Vol. 20, No. 9, December 2002.

Bibliography

Al-Jazzar, S., and J. Caffery, Jr., "A New Joint AOA/Delay Estimator for Wideband Spread Spectrum Systems," *1st Mobile and Wireless Commun. Summit*, Dresden, Germany, June 19-23, 2005.

Barcelo, F., L. de Nardis, and P. Tome, "Advances in Indoor Location," *LIAISON—ISHTAR Workshop*, Athens, Greece, September 28–29, 2006.

Borkowski, J., and J. Lempiainen, "Practical Network-Based Techniques for Mobile Positioning in UMTS," *EURASIP Journal on Applied Signal Processing*, 2006.

Botteron, C., M. Fattouche, and A. Host-Madsen, "Statistical Theory of the Effects of Radio Location System Design Parameters on the Positioning Performance," *Proc. Vehicular Technology Conference*, 2002.

Caffery, Jr., J. J., *Wireless Location in CDMA Cellular Radio Systems*, Norwell, MA: Kluwer Academic Publishers, 1999.

Cheney, M., "The Linear Sampling Method and the MUSIC Algorithm," Department of Electromagnetic Theory, Lund University, Lund, Sweden, February 7, 2003.

Cheng, C., and A. Sahai, "Cramer-Rao-Type Bounds for Locatization," *EURASIP Journal on Applied Signal Processing*, 2006.

Cheng, X., et al., "TPS: A Time-Based Positioning Scheme for Outdoor Wireless Sensor Networks," *IEEE INFOCOM*, March 7–11, 2004.

Chung, W. C., and D. S. Ha, "An Accurate Ultra Wideband (UWB) Ranging for Precision Asset Location," *IEEE Conference on Ultra Wideband Systems and Technologies*, November 16–19, 2003.

Ciurana, M., F. Barcelo, and S. Cugno, "Indoor Tracking in WLAN Location with TOA Measurements," *MOBIWAC*, Terromolinos, Spain, 2006.

Corral, C. A., S. Emami, and G. Rasor, "Ultra-Wideband Peak and Average Power Limits," *IEEE Consumer Communications and Networking Conference*, January 8–10, 2006.

Cyganski, D., et al., "Performance Limitations of a Precision Indoor Positioning System Using a Multi-Carrier Approach," *Institute of Navigation, National Technical Meeting*, San Diego, CA, January 26, 2005.

Falsi, C., et al., "Time of Arrival Estimation for UWB Localizers in Realistic Environments," *EURASIP Journal on Applied Signal Processing*, 2006.

Fontana, R. J., E. Richley, and J. Barney, "Commercialization of an Ultra Wideband Precision Asset Location System," *IEEE Conference on Ultra Wideband Systems and Technologies*, November 16–19, 2003.

Gao, Y., J. F. McLellan, and J. B. Schleppe, "An Optimized GPS Carrier Phase Ambiguity Search Method Focusing on Speed and Reliability," *IEEE AES Systems Magazine*, December 1996.

Gezici, S., et al., "Localization Via Ultra-Wideband Radios," *IEEE Signal Processing Magazine*, July 2005.

Goud, P., A. Sesay, and M. Fattouche, "A Spread Spectrum Radiolocation Technique and Its Application to Cellular Radio," *IEEE Pacific Rim Conference on Communications, Computers and Signal Processing*, May 9–10, 1991.

Hassan, E. N. E. H., "Mobile Localization in a GSM Network," Ph.D. thesis, Ecole Nationale Superieure des Telecommunications, Paris, 1999.

Hatami, A., and K. Pahlavan, "In-Building Intruder Detection for WLAN Access," *Position Location and Navigation Symposium*, April 26–29, 2004.

Ingram, S. J., D. Harmer, and M. Quinlan, "Ultra WideBand Indoor Positioning Systems and Their Use in Emergencies," *Position Location and Navigation Symposium*, April 26–29, 2004.

Kaemarungsi, K., and P. Krishnamurthy, "Modeling of Indoor Positioning Systems Based on Location Fingerprinting," *IEEE INFOCOM*, March 7–11, 2004.

Kennedy, J., and M. C. Sullivan, "Direction Finding and 'Smart Antennas' Using Software Radio Architectures," *IEEE Communications Magazine*, May 1995.

Kikuchi, S., A. Sano, and H. Tsuji, "Blind Mobile Positioning in Urban Environment Based on Ray-Tracing Analysis," *EURASIP Journal on Applied Signal Processing*, 2006.

Kolodziej, K. W., and J. Huelm, *Local Positioning Systems: LBS Applications and Services*, Boca Raton, FL: CRC Press, 2006.

Krim, H., and M. Viberg, "Two Decades of Array Signal Processing Research," *IEEE Signal Processing Magazine*, July 1996.

Kupper, A., *Location-Based Services: Fundamentals and Applications*, New York: John Wiley & Sons, 2005.

Liberti, J. C., and T. S. Rappaport, *Smart Antennas for Wireless Communications: IS-95 and Third Generation CDMA Applications*, Upper Saddle River, NJ: Prentice-Hall, 1999.

Maggio, G. M., "An Introduction to IEEE 802.15.4a," *MICS Meeting*, Neuchatel, February 21, 2006.

Misra, P., B. P. Burke, and M. M. Pratt, "GPS Performance in Navigation," *Proc. IEEE*, Vol. 87, No. 1, January 1999.

Nardis, L. D., and M. D. Benedetto, "Overview of the IEEE 802.15.4/4a Standards for Low Data Rate Wireless Personal Data Networks," *IEEE 4th Workshop on Positioning, Navigation and Communications 2007*, Hannover, Germany, 2007.

Pahlavan, K., P. Krishnamurthy, and J. Beneat, "Wideband Radio Propagation Medeling for Indoor Geolocation Applications," *IEEE Communications Magazine*, April 1998.

Patwari, N., et al., "Relative Location Estimation in Wireless Sensor Networks," *IEEE Trans. on Signal Processing*, Vol. 51, No. 8, August 2003.

Pickholtz, R. L., D. L. Schilling, and L. B. Milstein, "Theory of Spread-Spectrum Communications—A Tutorial," *IEEE Trans. on Communications*, Vol. COM-30, No. 5, May 1982.

Pirinen, T. W., "A Lattice Viewpoint for Direction of Arrival Estimation from Quantized Time Differences of Arrival," *4th IEEE Workshop on Sensor Array and Multichannel Processing*, Waltham, MA, July 12–14, 2006, pp. 50–54.

Progri, I. F., W. R. Michalson, and M. C. Bromberg, "Accurate Synchronization Using a Full Duplex DSSS Channel," *Position Location and Navigation Symposium*, April 26–29, 2004.

Roy, R., and T. Kailath, "ESPRIT—Estimation of Signal Parameters Via Rotational Invariance Techniques," *IEEE Trans. on Acoustics, Speech, and Signal Processing*, Vol. 37, No. 7, July 1989.

Roy, S., et al., "Ultrawideband Radio Design: The Promise of High-Speed, Short-Range Wireless Connectivity," *Proc. IEEE*, Vol. 92, No. 2, February 2004.

Saeed, R. A., et al., "Performance of Ultra-Wideband Time-of Arrival Estimation Enhanced with Synchronization Scheme," *ECTI Trans. on Electrical Eng., Electronics, and Communications*, Vol. 4, No. 1, February 2006.

Sayrafian-Pour, K. and D. Kaspar, "Application of Beamforming in Wireless Location Estimation," *EURASIP Journal on Applied Signal Processing*, 2006.

Schantz, H. G., "Introduction to Ultra-Wideband Antennas," *Conference on UWB Systems and Technologies*, Reston, VA, November 16–19, 2003.

Schmidt, R. O., "Multiple Emitter Location and Signal Parameter Estimation," *IEEE Trans. on Antennas and Propagation*, Vol. AP-34, No. 3, March 1986.

Shimizu, Y., and Y. Sanada, "Accuracy of Relative Distance Measurement with Ultra Wideband System," *IEEE Conference on Ultra Wideband Systems and Technologies*, November 16–19, 2003.

So, H. C., and F. K. W. Chan, "A Novel Signal Subspace Approach for Mobile Positioning with Time-of-Arrival Measurements," *EUSIPCO*, 2006.

So, H. C., and E. M. K. Shiu, "Performance of TOA-AOA Hybrid Mobile Location," *IEICE Trans. Fundamentals*, Vol. E86-A, No. 8, August 2003.

Stoica, L., et al., "An Ultrawideband System Architecture for Tag Based Wireless Sensor Networks," *IEEE Trans. on Vehicular Technology*, Vol. 54, No. 5, September 2005.

Stoica, P., and A. Nehorai, "MUSIC, Maximum Likelihood, and Cramer-Rao Bound," *IEEE Trans. on Acoustics, Speech, and Signal Processing*, Vol. 37, No. 5, May 1989.

Stoica, P., and A. Nehorai, "MUSIC, Maximum Likelihood, and Cramer-Rao Bound: Further Results and Comparisons," *IEEE Trans. on Acoustics, Speech, and Signal Processing*, Vol. 38, No. 12, December 1990.

Strom, E. G., et al., "Propagation Delay Estimation in Asynchronous Direct-Sequence Code-Division Multiple Access Systems," *IEEE Trans. on Communications*, Vol. 44, No. 1, January 1996.

Tekinay, S., E. Chao, and R. Richton, "Performance Benchmarking for Wireless Location Systems," *IEEE Communications Magazine*, April 1998.

Voltz, P. J., and D. Hernandez, "Maximum Likelihood Time of Arrival Estimation for Real-Time Physical Location Tracking of 802.11a/g Mobile Stations in Indoor Environments," *Position Location and Navigation Symposium*, April 26–29, 2004.

Wang, S. S., M. Green, and M. Malkawi, "Mobile Positioning Technologies and Location Services," *IEEE RAWCON*, 2002.

Xu, P., R. J. Palmer, Y. Jiang, "An Algorithm to Determine Range Distance with a Frequency Hopping Spread Spectrum System," *Canadian Conference on Electrical and Computer Engineering 2005*, Saskatoon, May 2005.

Xu, P., R. J. Palmer, Y. Jiang, "An Analysis of Multipath for Frequency Hopping Spread Spectrum Ranging," *Canadian Conference on Electrical and Computer Engineering 2005*, Saskatoon, May 2005, pp. 2135–2138.

Zagami, J. M., S. A. Parl, and J. J. Bussgang, "Providing Universal Location Services Using a Wireless E911 Location Network," *IEEE Communications Magazine*, April 1998.

List of Acronyms and Abbreviations

ACK	Acknowledge
A-D	Analog to digital
AGC	Automatic gain control
AGPS	Assisted GPS
AMPS	Advanced mobile phone system
AOA	Angle of arrival
AP	Access point
ARRL	American Radio Relay League
AWGN	Additive white Gaussian noise
BPF	Bandpass filter
BPSK	Binary phase shift keying
BS	Base station
C/A	Course acquisition
CDMA	Code division multiple access
CN	Core network
CPE	Customer premises equipment
CSMA/CA	Carrier sense multiple access with collision avoidance
CW	Continuous wave
DCM	Database correlation method
DDP	Dominant direct path
DF	Direction finding
DGPS	Differential GPS
DLL	Delay lock loop
DM	Distance measuring
DME	Distance measuring equipment
DOA	Direction of arrival
DOP	Dilution of precision
DSRC	Dedicated short-range communication
DSSS	Direct sequence spread spectrum
ECEF	Earth centered, Earth fixed
ED	Energy detection
EGNOS	Euro geostationary navigation overlay service
EIRP	Equivalent isotropic radiated power

E-OTD	Enhanced observed time differences
ESPAR	Electrically steerable parasitic array radiator
ESPRIT	Estimation of signal parameters via rotational invariance techniques
EV/FBCM	Eigenvector forward backward correlation matrix
FCC	Federal Communications Commission
FDMA	Frequency division multiple access
FFD	Full function device
FFT	Fast Fourier transform
FHSS	Frequency hopping spread spectrum
FOM	Figure of merit
GDOP	Geometric dilution of precision
GMSK	Gaussian minimum shift keying
GPRS	General packet radio service
GPS	Global positioning system
GRI	Group repetition interval
GSM	Global system for mobile communication
HCPR	Half-cycle peak ratio
HDOP	Horizontal dilution of precision
HF	High frequency
HOW	Handover word
IEEE	Institute of Electrical and Electronics Engineers
IF	Intermediate frequency
IFFT	Inverse fast Fourier transform
IFT	Inverse Fourier transform
IR-UWB	Impulse radio ultrawideband
ISDN	Integrated services digital network
ISI	Intersymbol interference
ISM	Industrial, scientific, medical
ITS	Intelligent transport systems
LAN	Local area network
LBS	Location-based services
LC	Location coordinator
LFSR	Linear feedback shift register
LMU	Location measurement unit
LOS	Line of sight
LPF	Lowpass filter
LQI	Link quality indication
MAC	Medium access control
MB-OFDM	Multiband orthogonal freqency division multiplex
MLC	Mobile location center
MP	Matrix pencil

MS	Mobile station
MSC	Mobile switching center
MSAS	Multifunctional satellite augmentation system
MUSIC	Multiple signal classification
NCO	Numerically controlled oscillator
NDDP	Nondominant direct path
NLOS	Nonline of sight
OFDM	Orthogonal frequency division multiplex
OOK	On-off keying
OTDOA-IPDL	Observed TDOA-idle period downlink
PAN	Personal area network
PCS	Personal communication system
PDN	Packet data network
PHY	Physical layer
PLL	Phase locked loop
PN, PRN	Pseudorandom noise
PPM	Pulse position modulation
PPS	Precise positioning service
PSAP	Public safety answering point
PSK	Phase shift keying
PSTN	Public switched telephone network
QPSK	Quadrature phase shift keying
RFD	Reduced function device
RFID	Radio frequency identification
RNC	Radio network controller
RSS	Received signal strength
RSSI	Receiver signal strength indication
RTD	Relative time differences
RTLS	Real-time location systems
RTT	Round trip time
SAS	Stand-alone serving mobile location center
SAW	Surface acoustic wave
SFD	Start frame deliminator
SIFS	Short interframe space
SMLC	Serving mobile location center
SNR	Signal-to-noise ratio
SPS	Standard positioning service
SUI	Stanford University interim (model)
TA	Timing advance
TDMA	Time division multiple access
TDOA	Time difference of arrival
TDOP	Time dilution of precision

TLM	Telemetry word
TOA	Time of arrival
TOF	Time of flight
UDP	Undetected direct path
UE	User equipment
UMTS	Universal mobile telecommunication system
USDC	U.S. digital cellular standard
UTC	Universal coordinated time
U-TDOA	Uplink time difference of arrival
UWB	Ultrawideband
VCO	Voltage controlled oscillator
VCXO	Voltage controlled crystal oscillator
VDOP	Vertical dilution of precision
VOR	Very high frequency omnidirectional ranging
WAAS	Wide area augmentation system
WLAN	Wireless local area network
WPAN	Wireless personal area network

About the Author

Alan Bensky is an electronics engineering consultant with over 25 years of experience in analog and digital design, management, and marketing. Specializing in wireless circuits and systems, Mr. Bensky has carried out projects for varied military and consumer applications and led the development of three patents on wireless distance measurement. He is the author of *Short-Range Wireless Communication, Second Edition* (Elsevier, 2004), and has written several articles in international and local publications. He has taught courses and given lectures on radio engineering topics. Mr. Bensky is a senior member of the IEEE.

Index

A

Access point (AP), 148–58, 243–49
Accuracy, defined, 4
ACK (acknowledge), 243–48
Acquisition
  AOA, 194
  DME, 36–37
  multicarrier phase, 124
  spread spectrum, 63–71, 82–92
  UWB, 278, 284
  See also C/A (course acquisition)
Adaptive array, 209
Adaptive controller, 210
Adaptive equalizer, 119
Additive white Gaussian noise (AWGN), 112, 278
Advanced mobile phone system (AMPS), 237, 238
Alzheimer patients, 5
Ambiguity
  AOA, 195, 199, 204–5
  carrier phase, 93, 108
  intersecting circles, 9, 44
  measurement delay, spread spectrum, 80
  range, spread spectrum, 87–88
  rho-rho method, 29
  target direction, VOR, 38
  TOA, 163, 170
American Radio Relay League (ARRL), 192
Amplitude comparison, 198
Angle of arrival (AOA), 189–221
  cellular networks, 223, 236, 237
  measurement, basic, 2
  overview, 9
  RFID, 260–62

  theta-theta, 28
  VOR, 38
Angulation, defined, 3
Animal tracking, 5, 29
  AOA, 189, 196, 205–6
  cellular network, 223
  See also Wildlife tracking
Antenna directivity, defined, 192
Antenna gain, defined, 192
Antenna pattern, defined, 192
Asset tracking, 5
Assisted GPS (AGPS), 231
Authentication routine, 260
Autocorrelation, 57–58, 279
  See also Correlation, Cross correlation
Automatic gain control (AGC), 82, 85

B

Bandwidth
  chirp signal, 22–24
  effect on time resolution, 15–17
  effective, 19–20, 278
  fractional (UWB), 266
  GPS, 49
  IEEE 802.15.4a, 276
  spread spectrum, 53–56
Barker codes, 24
Base station (BS), defined, 3
Bayesian inference, 149, 153–58
Beacon
  defined, 3,
  DME, 33–37
  person and asset tracking, 5
  rho-rho, 29
  RSS, 146
  TDOA, 83, 105, 173
  VOR, 37–38

Beamforming, 220
Beam width, 37, 190–200, 210
    defined, 192
Binary phase shift keying (BPSK)
    carrier acquisition, 69
    carrier tracking, 72–73
    OFDM modulation, 129
    spread spectrum modulation, 54–56,
        82
    UWB modulation, 273
Bluetooth, 119, 251–54
Broadside antenna, defined, 194

C
Cardiod antenna radiation pattern, 196
Carrier acquisition, 64
Carrier phase ranging, 92
Carrier sense multiple access with
        collision avoidance (CSMA/CA),
        243
Carrier synchronization, 49, 61–63
Carrier tracking, 72
CDMA IS-95, 225
Cell-ID (cell identification), 228–31
Cellular networks, 223–40
Challenge-response dialog, 260
Chip, defined, 24
Chirp modulation
    IEEE 802.15.4a, 274
    pulse compression, 20–22, 24
Closed form solution, 164
Clock bias, 45
Clock jitter, 181
Coarse acquisition (C/A), 45, 71, 87
    accuracy, 93
    Gold codes, 60–61
Cochannel interference, 51, 53, 186, 237
Code acquisition, 64
Code despreading, 62
Code division multiple access (CDMA)
    cellular networks, 227, 225–28
    cellular positioning problems, 237
    cochannel interference, 186
    Galileo, 51
    GPS, 45
    UMTS 235
    WAAS, 50

Code tracking, 72
Coordinate systems, 6
Correlation
    basic principles, 24–27
    cellular network, 237
    code acquisition, 69–70
    code synchronization, 61–62
    code tracking, 73–76
    despreading, 57–58
    direction finding, 194
    matrix, 185
    multipath, 90–91
    range gate, 35
    receiver, UWB, 270–72
    TDOA, 233
    See also Autocorrelation, Cross
        correlation
Core network (CN), 228
Costas loop, 72, 73
Cramer-Rao inequality, 182, 278
Cross correlation
    cellular network, equation, 233
    definition and qualities, 58–59
    GPS, 49
    IEEE 802.15.4a, 276, 279
    TDOA, equations, 172–73
    See also Autocorrelation, Correlation
Cross power spectral density, 172–73
Customer premises equipment (CPE),
        145

D
Database correlation method (DCM),
        237
Dedicated short-range communication
        (DSRC), 6
Delay lock loop (DLL), 73, 75, 89–92
Deliminator. See SFD
Delta impulse function, 185
Delta pseudorange, 63
Despreading, 27, 53–63, 71
Differential GPS (DGPS), 50
Digital correlator, 26
Dilution of precision (DOP), 182–84. See
        also GDOP, HDOP, VDOP,
        TDOP
Dipole antenna, 192, 195, 209

Direct sequence spread spectrum (DSSS), 53–62
Direction of arrival (DOA). *See* AOA
Direction finding (DF), 27, 189
Discriminator types, 89
Distance bounding, 9, 272, 275
Distance measuring (DM),
  basic properties, 2–4
  overview, 6
Distance measuring equipment (DME), 33–37
Diversity reception, 131
Dominant direct path (DDP), 184, 185
Doppler shift, 62–63, 70, 119
Downlink, defined, 237
Duplex mode, 79
Duty cycle, 15, 267, 272, 284

E

E-112, 4, 240
E-911, 4, 239
Earth centered, Earth fixed (ECEF), 46
Eavesdropping, 283
Eigenvector forward backward
  correlation matrix (EV/FBCM), 209
Electrically steerable parasitic array
  radiator (ESPAR), 238–45
Electronically steered antennas, 29, 207
Encryption, 46, 260, 283
End-fire array, 194, 195
Energy detection (ED), 256
Enhanced observed time differences
  (E-OTD), 258, 259
Epoch, defined, 4
Equivalent weight vector, 218
Estimation of signal parameters via
  rotational invariance techniques
  (ESPRIT), 209, 314
Euclidian distance, 149, 157, 158
Euro geostationary navigation overlay
  service (EGNOS), 50

F

Fast Fourier transform (FFT), 128–36, 280

Federal Communications Commission
  (FCC)
  cellular location, 4, 224
  unlicensed band, 257
  UWB, 265–67
Figure of merit (FOM), 281, 282
Fingerprinting, 2, 10
  RSS, 139, 147, 241, 249
Fleet management, 6, 223
Forward channel, 225
Free space, 9, 140–43, 257
Frequency division multiple access
  (FDMA), 51
Frequency hopping spread spectrum
  (FHSS), 110, 122, 125, 251
Frequency domain cross correlation, 173
Full function device (FFD), 255

G

Galileo, 45, 50–51
Gaussian minimum shift keying (GMSK), 226
General packet radio service (GPRS), 230
Geolocation, 286
Geometric dilution of precision (GDOP),
  153, 183, 239. *See also* DOP
Global positioning system (GPS), 44–51,
  60–63, 70–71
  cellular location, 228–31, 238, 239
  constellation, 46
  frequencies of operation, 45
  introduction, 1, 4, 5, 10
  performance, 181–86
  TOA, 166–71
Global system for mobile communication
  (GSM), 225–28
  E-OTD, 234
  problems, 238
  timing advance, 232
Glonass, 45, 50–52
Gold codes, 59, 60
Golden received power range, 253
Gradient descent, 164
Group repetition interval (GRI), 42

H

Half-cycle peak ratio (HCPR), 44
Half duplex, 79, 83

Handoff, 5, 186
Handover word (HOW), 48
Handset based, 229
Helical antenna, 193
High frequency (HF), 189
Histograms, 155, 167–69
Horizontal dilution of precision (HDOP),
        183
Horn antenna, 196, 198
Hyperbola
    basics, 31–32
    cellular TDOA, 227–28
    Loran-C, 41–42
    phase difference, 125
    TDOA in WLAN, 248
    TDOA model, 175–79
Hyperbolic curves, 30. *See also*
        Hyperbola
Hyperboloid, 125. *See also* Hyperbola
Hybrid positioning system, 230

I

IEEE (Institute of Electrical and
        Electronics Engineers)
    IEEE 802.11, 242–44, 249–51
    IEEE 802.15.1. *See* Bluetooth
    IEEE 802.15.3a, 258
    IEEE 802.15.4, 255
    See IEEE 802.15.4a
    IEEE 802.16, 144
IEEE 802.15.4a, 251, 257, 274–83
    crystal offset correction, 282
    ranging protocol, 280
    transmit spectral mask, 278
Impulse radio ultrawideband (IR-UWB),
        268–72
    coherent receivers, 270–72
    noncoherent receivers, 270–72
Industrial, scientific, medical (ISM), 125
Integrated services digital network
        (ISDN), 225
Intelligent transport systems (ITS), 223
Intersymbol interference (ISI), 276
Inverse fast Fourier transform (IFFT),
        128, 273
Inverse Fourier transform (IFT), 185

Isotropic antenna, 192
Isotropic gain, 192

J

Jamming, 44, 53, 88

L

Lateration, defined, 3
Least-mean-square (LMS), 170, 214
Least squares (LS), error criterion, 164,
        165
Linear feedback shift register (LFSR), 59,
        68, 83
Line of sight (LOS), 271, 283–86
Link quality indication (LQI), 255, 256
Local area network (LAN). *See* WLAN
Location awareness, 5, 11, 139, 146, 147
Location-based services (LBS), 4, 230
Location coordinator (LC), 149
Location, defined, 3
Location measurement unit (LMU), 225,
        245–47
Loran-C, 38–44
    master station, 40, 42
    secondary stations, 40, 42

M

M-ary phase shift keying, 54, 272
Matched filter, 20–27, 61–62, 71, 85–86,
        102–3
Matrix pencil (MP), 185
Medium access control (MAC), 103,
        154, 258, 259, 274
Mobile location center (MLC), 234
Mobile station (MS), defined, 3
Mobile switching center (MSC), 225
M-sequence codes, 59–60, 64–65, 74
Multiband orthogonal frequency division
        multiplex (MB-OFDM), 268, 274
Multicarrier distance measuring, 107,
        123, 125, 126
    See also Multicarrier phase
        measurement
Multicarrier phase measurement, 107–37
Multifunctional satellite augmentation
        system (MSAS), 50
Multilateral system, defined, 4

Multipath, 16, 17, 77
    AOA, 196, 206, 221
    cellular system, 228, 235, 237, 239
    fingerprinting, 249
    multicarrier phase, 118–22, 125
    OFDM, 127, 128, 133, 134
    RFID, 241, 242, 247
    RSS, 139, 143, 161
    spread spectrum, 88, 90–92
    time transfer, 102
    TOA, 163, 184–87
    UWB, 265, 270–74, 276, 278, 282–86
    WLAN, 250, 251
    WPAN, 255–57
Multiple signal classification (MUSIC),
    185, 205, 206

N
Navigation, defined, 4
Navstar. *See* GPS
Near-far problem, 186, 228
Network allocation vector (NAV), 243
Network security, 6
Newton's method, 170, 214
Noise, 17–20
    random, 182
    range error, 101
    tracking, 88
Noise bandwidth, 16
Nondominant direct path (NDDP), 184,
    185
Nonline-of-sight (NLOS), 184, 185
Numerically controlled oscillator (NCO),
    76, 77, 82–86

O
Observed TDOA (OTDOA), 229, 235
Observed TDOA-idle period downlink
    (OTDOA-IPDL), 228
OFDM distance measurement, 130–34
On-off keying (OOK), 284
Open field propagation, 141
Orthogonal frequency division multiplex
    (OFDM), 126–37
Overdetermined
    equations, 163–66
    system, 177

P
Packet data network (PDN), 225
Pan-European cellular system. *See also*
    GSM
Pattern recognition, 147, 236, 237
P-code, 45, 46
Personal area network (PAN). *See*
    WPAN
Personal communication system (PCS),
    224
Phase detector, 38, 112, 121, 200–7
Phase difference measurements, 125
Phase integrator, 83, 84
Phase interferometer, 198, 200–9, 262
Phase locked loop (PLL), 72–73, 124–25,
    203
Phase shift keying (PSK), 72, 82
Phase slope method, 108
Physical layer (PHY)
    ECMA-368, 103, 258
    IEEE 802.15.4a, 279, 280
Precise positioning service (PPS), 49
Precision, defined, 4
PRN. *See* PN
Propagation time, 2
    distance bounding, 261
    ECMA-368, 259
    effect of relative motion, 102
    multicarrier phase slope, 115
    multipath, 118, 121–22
    OFDM, 130, 134, 136
    resolution, 88–92
    time transfer, 96–97, 99–101
    timing advance, 226
    TOA estimate, 278
    types, 77
    WLAN, 244–45
Proximity
    defined, 2
    RFID, 260
Pseudorandom noise (PN), 59
Pseudorandom sequence, 4, 51, 57, 119,
    125, 269
Pseudorange, 50, 63, 78, 83, 167, 231
Public safety answering point (PSAP),
    224

Public switched telephone network
  (PSTN), 225
Pulse compression, 20, 24
Pulse position modulation (PPM),
  268–71
Pulse repetition rate, 15, 20, 36
Pulse rise time, 15, 17
Polarization, defined, 193

Q
Quadrature phase shift keying (QPSK),
  73, 82, 128, 273

R
Radar, 6–8, 13–15, 171, 189
  RFID, 262
  secondary radar, 33
Radio frequency identification (RFID),
  259–62
Radiometric detection, 71
Radiometer, 71
Radio network controller (RNC), 228,
  229, 235
Radio wave propagation, 6
Rake receiver, 88, 91, 119
Range,
  bias, 169
  defined, 3
  gate, 35–37
Ray tracing, 139, 148, 150, 151
Real-time location systems (RTLS), 259
Received signal strength (RSS), 139–58
  Bayesian inference, 153–57
  cellular networks, 236
  database search method, 147
  nearest neighbor, 149–53, 157
  smart sensor distributed networks, 147
Reduced function device (RFD), 255
Relative time differences (RTD), 235
Remote positioning system, 230
Replica code, 57–64
Resolution, defined, 4
Reverse channel, 225
Rho-rho, 29
Rho-theta, 27
Round-trip time (RTT), 228
Rogue terminal, 6, 140, 250

RSSI, 139, 157
  AOA, 194, 215
  Bluetooth, 252–69
  cellular, 236
Root raised cosine filter, 276
Root raised cosine pulse, 276

S
Satellite navigation, 45, 50, 186
Self-positioning, 230
Serving mobile location center (SMLC),
  229
Shift register, 26, 60, 84
Short interframe space (SIFS), 243–46
Sidelobes, 25, 58, 90, 210, 279
Sliding correlator, 61, 68, 88, 171
Smart antenna, 10, 208, 236, 251
Smart cards, 260, 261
Squaring loop, 672, 673
Stand-alone serving mobile location
  center (SAS), 228, 229
Standard positioning service (SPS), 49
Stanford University interim model (SUI),
  144
Start frame deliminator (SFD), 243, 270,
  271, 278–80
Station, defined, 3
Steering vector, 210, 219
Surface acoustic wave (SAW) delay line,
  24
Switched beam antenna, 208
Synchronization
  beacon, 3
  code and carrier, GPS, 49
  frame, 27
  GSM, 226–27
  IEEE 802.15.4a, 274, 278, 281
  spread spectrum
    acquisition, 63, 81
    carrier tracking, 69
    code, 66–69, 73
    code phase, 79
    frequency, 85
    principles, 60–62
  TDOA, 170–71, 181
  telemetry word, GPS, 48
  time transfer, 103

UWB, 270–72, 274, 284
TDOA, cellular system, 233
WLAN, 242–43

T

Tags, RFID, 260
Target, defined, 3
Taylor series expansion, 164
Telemetry word (TLM), 48
Terminal, defined, 3
Texas Instruments, 256
Theta-theta, 28
Time of arrival (TOA), 2, 29, 162–67, 242
Time bias, 170, 187
Time difference of arrival (TDOA), 2, 30, 170–80
  cellular, 232–36
  multicarrier phase, 125
  WLAN, 248
Time dilution of precision (TDOP), 184
Time division multiple access (TDMA), 226, 237–39
Time of flight (TOF), defined, 2
Time hopping, 53, 268–69, 284
Time transfer, 95–105
  calibration, 97, 98, 103
  relative motion, 102
Timestamp
  cellular networks, 229, 237
  ECMA-368, 258
  elapsed time measurement, 78–83
  fingerprinting, 249
  resolution in 802.11, 246–47
  time transfer, 95
  TOA basics, 30
Timing advance (TA), 226–28, 232
Timing synchronization function (TSF), 242
Toll station, 6
Traffic telematics, 6
Triangulation, defined, 3
Trilateration, defined, 3

U

Ultrawideband (UWB), 265–86
  fractional bandwidth, 266

wall and ground penetration imaging, 276
Undetected direct path (UDP), 184
Unilateral system, defined, 4
Universal coordinated time (UTC), 45, 48
Universal mobile telecommunication system (UMTS), 225, 228, 235
Uplink, 225
Uplink time difference of arrival (U-TDOA), 228, 229, 236
User equipment (UE), 228
U.S. digital cellular standard (USDC), 237

V

Velocity, 3, 45, 48, 62, 63
Vertical dilution of precision (VDOP), 84
Very high frequency omnidirectional ranging (VOR), 28, 33, 37–38
Voltage controlled oscillator (VCO), 68–69, 72–76, 123, 125

W

WCDMA, 275
Wide area augmentation system (WAAS), 50
WiFi. *See* WLAN
Wildlife tracking, 5, 28. *See also* Animal tracking
Wireless handcuffs, 5
Wireless local area network (WLAN), 242–49
Wireless personal area network (WPAN), 251–58

Y

Yagi array, 194, 196–97, 215

Z

ZigBee, 255–57

The GNSS Technology and Applications Series

Elliott Kaplan and Christopher Hegarty, Series Editors

Applied Satellite Navigation Using GPS, GALILEO, and Augmentation Systems, Ramjee Prasad and Marina Ruggieri

Digital Terrain Modeling: Acquisition, Manipulation, and Applications, Naser El-Sheimy, Caterina Valeo, and Ayman Habib

Geographical Information Systems Demystified, Stephen R. Galati

GNSS Markets and Applications, Len Jacobson

GNSS Receivers for Weak Signals, Nesreen I. Ziedan

Introduction to GPS: The Global Positioning System, Second Edition, Ahmed El-Rabbany

Principles of GNSS, Inertial, and Multisensor Integrated Navigation Systems, Paul D. Groves

Spread Spectrum Systems for GNSS and Wireless Communications, Jack K. Holmes

Understanding GPS: Principles and Applications, Second Edition, Elliott Kaplan and Christopher Hegarty, editors

Wireless Positioning Technologies and Applications, Alan Bensky

For further information on these and other Artech House titles, including previously considered out-of-print books now available through our In-Print-Forever® (IPF®) program, contact:

Artech House Publishers
685 Canton Street
Norwood, MA 02062
Phone: 781-769-9750
Fax: 781-769-6334
e-mail: artech@artechhouse.com

Artech House Books
46 Gillingham Street
London SW1V 1AH UK
Phone: +44 (0)20 7596 8750
Fax: +44 (0)20 7630 0166
e-mail: artech-uk@artechhouse.com

Find us on the World Wide Web at: www.artechhouse.com